Studies in Rhythm Engineering

This is a multi-disciplinary book series, ranging from astrophysics to biology, chemistry, mathematics, geophysics and materials science. Its primary scope is the fundamental science and associated engineering wherever cyclic and rhythmic oscillations are observed.

Time neither being an entity nor a process is unmeasurable and undefined, although a clock only measures the passage of time. The clock drove recurring processes are observed in the biological rhythms, astrophysical and geophysical environments. Always, the clocks are nested, arranged in a geometric shape to govern a phenomenon in nature. These clocks are made of atoms, molecules, their circuits, and complex networks. From biology to astrophysics, the clocks have enriched the science and engineering and the series would act as a catalyst and capture the forthcoming revolution of time cycles expected to unfold in the 21st Century. From the cyclic universe to the time crystal, the book series makes a journey through time to explore the path that time follows.

The series publishes monographs and edited volumes.

Aditi Kothari Chhajer · Vinod Singh Gour
Editors

Introduction to Rhythms in Plants

Springer

Editors
Aditi Kothari Chhajer
Department of Botany
Sri Venkateswara College
Delhi University
New Delhi, Delhi, India

Vinod Singh Gour
Department of Education in Science
and Mathematics
NCERT-Regional Institute of Education
Mysuru, Karnataka, India

ISSN 2524-5546 ISSN 2524-5554 (electronic)
Studies in Rhythm Engineering
ISBN 978-981-95-3591-0 ISBN 978-981-95-3592-7 (eBook)
https://doi.org/10.1007/978-981-95-3592-7

This Springer imprint is published by the registered company Springer Nature Singapore Pte Ltd.
The registered company address is: 152 Beach Road, #21-01/04 Gateway East, Singapore 189721, Singapore

To
My father, whose love, strength, inspiration, and quiet wisdom gave the perfect rhythm to my life.

—Dr. Aditi Kothari Chhajer

To
My beloved mother who inculcated the values, ethics, sincerity, and hardworking traits in my life.

—Vinod Singh Gour

Preface

Plants, often perceived as passive and static organisms, in fact display a remarkable array of rhythmic behaviours that reveal their deep and dynamic connection with the environment. These rhythms are far from incidental; they are central to optimizing plant function, enhancing survival, and supporting growth and development. The rhythm of plant life is perhaps most vividly illustrated in seasonal phenomena—like the breathtaking bloom of cherry blossoms that captivates us each spring. Yet, this beauty is not a constant display. It arrives at a specific time of year, raising the question: *How do plants know when it's time?*

Plants possess an innate ability to perceive environmental cues—annually, seasonally, daily, and even hourly—and to respond in precisely regulated patterns. Among the most well-known of these behaviors are circadian rhythms: internal time-keeping mechanisms that allow plants to anticipate and adapt to the daily cycle of light and darkness. Operating on a near 24-hour cycle, these rhythms orchestrate a wide range of physiological processes, from leaf and flower movement to metabolic activities. Light-regulated processes such as photoperiodism, photosynthesis, and photomorphogenesis further underscore the vital relationship between plants and light. These responses are mediated by complex genetic networks and plant hormones, which are finely tuned by environmental signals.

The study of plant rhythms is more than an academic endeavour, it offers profound insights into plant physiology and ecology, and by extension, into broader environmental interactions. From circadian rhythms and photoperiodism to sleep and tropical movements, these behaviours reflect the astonishing adaptability and resilience of plants. Starting with Chap. 1 which gives an insight into the rhythmic behaviour of plants, the book traverses through various landscapes like photomorphogenesis and photoperiodism, leaf and stomatal movements and finally to pollination as affected by the dynamic circadian cycle.

We believe this book will facilitate a deeper understanding of the inter-relationship between plant physiology, its genetics, the molecular network of genes and the internal time keeping clock. For the undergraduate and post-graduate student, this book will compel them to look at science holistically rather than being stuck up in water-tight boundaries. As they delve into this fascinating realm, they shall gain not only scientific understanding but also a renewed appreciation for the dynamic, responsive nature of the plant kingdom. This book invites readers to explore the intricate symphony of plant life, which ensures the survival and success of these extraordinary organisms.

New Delhi, India — Aditi Kothari Chhajer
Mysuru, India — Vinod Singh Gour

Acknowledgements

We extend our heartfelt gratitude to all the scientists, students, and academicians who have contributed to the field of plant rhythms through their research and publications. Their pioneering work has laid the foundation for our understanding and has been a constant source of inspiration in shaping this book. It is through their publications and insights that we were able to delve deeper into the fascinating world of rhythmic phenomena in plants.

Our special thanks to **Prof. S. L. Kothari** for guiding us throughout the journey and offering his invaluable words of wisdom and insights. We are especially thankful to **Prof. Kanad Ray** for providing us with this valuable opportunity and for his continued encouragement and support throughout the development of this work.

We also wish to acknowledge the unwavering support of the team at **Springer**. Our sincere thanks to **Mr. Aninda Bose**, Executive Editor, for his guidance and enthusiasm, and to **Ms. Divya Meiyazhagan**, Production Editor, for her meticulous attention to detail and coordination during the production process. Their professionalism and assistance have been instrumental in bringing this book to fruition. Last but not the least, we would like specially thank our families who have offered their constant support during the preparation of this book. This would not have been possible without their support.

Contents

1 **The Circadian Rhythms in Plants: An Introduction** ... 1
Aditi Kothari Chhajer, Pamil Tayal, Manoj Thakur, Gauri Jha, Shivam Saini, and S. L. Kothari

2 **Plant Response to Environmental Periodicity: Diurnal Rhythms** ... 31
Prasann Kumar, Shipa Rani Dey, Joginder Singh, and Neha Gogia

3 **Environmental Rhythms and Regulation of Plant Morphogenesis: Temporal Architectures—The Environmental Clocks of Plant Life** ... 71
Urszula Zajączkowska

4 **Circadian Regulation of Rhizomicrobiome** ... 81
Reema Kumar and Aditi Kothari Chhajer

5 **Circadian Rhythms in the Algal World** ... 95
Atreyee Bagchi, Deepanshu Vasuja, and Aditi Kothari Chhajer

6 **Rhythms of Leaf and Stomatal Movements** ... 125
Prasann Kumar, Shipa Rani Dey, Joginder Singh, and Neha Gogia

7 **Light and Plant Life: Photoperiodism and Photomorphogenesis** ... 161
Prasann Kumar, Amit Raj, Shipa Rani Dey, Joginder Singh, and Neha Gogia

8 **Cell Division in Plants is Influenced by Diurnal Rhythm** ... 201
Prasann Kumar, Shipa Rani Dey, Joginder Singh, and Neha Gogia

9 **The Pulse of Pollination: Bees and Their Role in Plant Reproductive Rhythms** ... 231
Abhishek Rajpurohit, Seema Yadav, and Manish Sarwa

About the Editors

Dr. Aditi Kothari Chhajer is an Assistant Professor at Sri Venkateswara College, Delhi University. She has been actively involved in teaching plant physiology, genetic engineering, economic botany and intellectual property rights at undergraduate level for the last 14 years. As an academic, she is dedicated to nurturing the next generation of scientists and researchers, imparting her knowledge through both teaching and mentoring. Her research focus has been on tissue culture, abiotic stress tolerance and *Agrobacterium* mediated genetic transformation in plants, particularly cereals and millets. Her research work is published in journals of national and international repute. She has done part of her doctoral research at Rothamsted Research Institute, UK as a Commonwealth Split Site Doctoral Fellow. She has been awarded a doctorate in Botany from University of Rajasthan, Jaipur. She was also awarded the Sakura Science Fellowship (Japan).

Dr. Vinod Singh Gour has been awarded a doctorate in Forest Genetics from Forest Research Institute, University Dehradun (India) in 2009. He has teaching experience of 14 years at UG and PG levels. He has guided five doctorate students and more than 15 masters projects. His areas of expertise include plant tissue culture, genetic diversity of trees, applied microbiology and plant ecology. He has published more than 40 research and review articles in prestigious journals. Besides, he has published ten book chapters and one book. He handled projects funded by ICMR, SERB, DST (GoI), DST (GoR) and NCERT. He has been awarded CSIR Junior Research fellowship, Willey Research Academy award and Sakura Science Fellowship (Japan). He is a life time member of Society of Plant Biochemistry and Biotechnology (India).

Abbreviations

AANAT	Aralkylamine N-Acetyltransferase
ABA	Abscisic Acid
ABAR	ABA-Related
AGPase	ADP-Glucose Pyrophosphorylase
ARFs	Auxin Response Factors
AUX/IAA	Auxin/Indole-3-Acetic Acid
BRI1/BAK1	Brassinosteroid Insensitive 1/BRI1-Associated Receptor Kinase 1
BRs	Brassinosteroids
CAB	Chlorophyll A/B-Binding
CCA1	Circadian Clock Associated 1
CDKB1	Cyclin-Dependent Kinase B1
CDKs	Cyclin-Dependent Kinases
CEA	Controlled Environment Agriculture
CHS	Chalcone Synthase
CK	Cytokinins
CKIs	Cyclin-Dependent Kinase Inhibitors
CO	Constans
CRISPR-Cas9	Clustered Regularly Interspaced Short Palindromic Repeats—Associated Protein 9
Cry	Cryptochrome
Cry1/Cry2	Cryptochrome 1/2
CRY2	Cryptochrome 2
CYCA	Cyclin A
CYCD	Cyclin D
EIN2	Ethylene Insensitive 2
EIN3/EIL	Ethylene Insensitive 3/EIN3-Like
ELF3	Early Flowering 3
FAD	Flavin Adenine Dinucleotide
FT	Flowering Locus T
GAs	Gibberellins

GH	Growth Hormone
GI	Gigantea
GID1	Gibberellin Insensitive Dwarf 1
GWD	Glucan Water Dikinase
HSP70	Heat Shock Protein 70
HY5	Elongated Hypocotyl 5
IAA	Indole-3-Acetic Acid
ICK	Interactor/Inhibitor of Cyclin-Dependent Kinase
IGF-1	Insulin-Like Growth Factor 1
JA	Jasmonic Acid
KRP	Kip-Related Protein
LED	Light Emitting Diode
LHY	Late Elongated Hypocotyl
LOV	Light, Oxygen, or Voltage
LuxR/LuxI	Quorum sensing regulators in bacteria
MAD2	Mitotic Arrest Deficient 2
MAPK	Mitogen-Activated Protein Kinase
MCM	Minichromosome Maintenance
MT1/MT2	Melatonin Receptor 1/2
NCED	9-cis-Epoxycarotenoid Dioxygenase
NR	Nitrate Reductase
NRF2	Nuclear Factor Erythroid 2-Related Factor 2
NRT1.1	Nitrate Transporter 1.1
ORC1	Origin Recognition Complex 1
p21, p27	Cyclin-Dependent Kinase Inhibitors (animal-related)
PAL	Phenylalanine Ammonia-Lyase
PHR	Photolyase Homologous Region
Phy	Phytochrome
PhyA/PhyB	Phytochrome A/B
PHYB	Phytochrome B
PI3K/AKT/mTOR	Signaling pathway (common in animals, not plants)
PIFs	Phytochrome Interacting Factors
PRR	Pseudo-Response Regulator
PRR5	Pseudo-Response Regulator 5
PYR/PYL	Pyrabactin Resistance/Pyrabactin Resistance-Like
PYR/PYL/RCAR	ABA Receptors (Pyrabactin Resistance/PYR-Like/ Regulatory Component of ABA Receptor)
RBR	Retinoblastoma-Related Protein
ROS	Reactive Oxygen Species
SA	Salicylic Acid
SnRK2	Snf1-Related Protein Kinase 2
SOC1	Suppressor of Overexpression of Constans 1
TIR1	Transport Inhibitor Response 1
TOC1	Timing of CAB Expression 1
TOR	Target of Rapamycin

TPH1	Tryptophan Hydroxylase 1
UVR8	UV Resistance Locus 8
ZTL	Zeitlupe

Chapter 1
The Circadian Rhythms in Plants: An Introduction

Aditi Kothari Chhajer, Pamil Tayal, Manoj Thakur, Gauri Jha, Shivam Saini, and S. L. Kothari

1 Introduction

Imagine walking through a dense forest at dawn—the leaves of certain plants begin to unfurl, flowers open to greet pollinators, and the morning sun triggers a cascade of metabolic changes deep within plant cells. These movements and physiological shifts are not random; they are orchestrated by a sophisticated internal timekeeping system known as the circadian rhythm.

Unlike animals, which can migrate, hibernate, or seek shelter, plants remain rooted in place, fully exposed to environmental fluctuations. Their survival hinges on their ability to anticipate rather than merely react to changes in light, temperature, and resource availability. By doing so, plants optimize their biological processes to align with the most favorable conditions.

Some of the most crucial plant functions controlled by circadian rhythms include:

Photosynthesis → Initiates precisely at dawn, maximizing light energy capture. [72]

Metabolism and Energy Storage → Ensures sugars and starch are synthesized during the day and consumed at night [8, 51].

Growth and Development → Hypocotyl elongation, leaf expansion, and root growth are all regulated by circadian-controlled genes [72]

Flowering and Reproduction → Plants synchronize flowering time with seasonal photoperiods to align with pollinators and favorable conditions [30, 68].

A. K. Chhajer · P. Tayal · M. Thakur
Department of Botany, Sri Venkateswara College, Delhi University-South Campus, Delhi, India

G. Jha · S. Saini
Department of Biological Sciences, Sri Venkateswara College, Delhi University-South Campus, Delhi, India

S. L. Kothari (✉)
Amity University Rajasthan, Jaipur, India
e-mail: slkothari28@gmail.com

A. Kothari Chhajer and V. Singh Gour (eds.), *Introduction to Rhythms in Plants*, Studies in Rhythm Engineering, https://doi.org/10.1007/978-981-95-3592-7_1

Stress Responses → Circadian rhythms prepare plants for drought, cold, and pathogen attacks, increasing their chances of survival [44].

For plants, whose survival depends on synchronizing their activities with the availability of sunlight, an internal clock is not just a convenience—it is a matter of life and death. The opening and closing of flowers, the precise timing of photosynthesis, the careful regulation of energy reserves overnight, and even the ability to anticipate seasonal changes are all orchestrated by this unseen yet ever-present biological rhythm [46].

Unlike a mere reaction to the environment, circadian rhythms allow plants to predict changes before they occur. A sunflower, for instance, does not follow the sun reactively—it anticipates its movement, priming itself to maximize light capture before dawn even breaks [4]. Similarly, a crop plant regulates its starch metabolism overnight with careful precision, ensuring it does not deplete its energy reserves too early [8, 43, 46, 51].

However, circadian rhythms are not fixed—they must be reset daily, just as a traveler adjusts their watch when entering a new time zone. This synchronization, known as entrainment, aligns the plant's internal clock with environmental signals. This ensures that biological functions are always aligned with the best possible environmental conditions [9].

Self- Sustained, Yet Adaptable: The Dual Nature of Circadian Rhythms

One of the most remarkable aspects of circadian rhythms is their self-sustaining nature. Even when plants are placed in constant darkness, their internal clock continues to cycle with a period of approximately 24 h [15]. This points to a possibility that the circadian rhythm responses at genetic and metabolic level are anticipatory and, not merely a response to external stimuli.

However, circadian rhythms must also be entrained—or synchronized—with real-world environmental changes to remain accurate. This synchronization is achieved through three primary environmental cues:

Light—The most powerful zeitgeber ("time giver") for plants, detected by phytochromes and cryptochromes, resetting the clock at dawn and dusk. [10, 16, 69]

Temperature—A secondary entrainment cue, ensuring clock stability across different climates [15, 22, 61]

Metabolic Signals—Sugars, nitrogen compounds, and hormones like ABA (abscisic acid) and GA (gibberellins) provide feedback loops, linking metabolic rhythms to external conditions [46].

In the following sections, we will explore how plants integrate these external cues into their biological clocks, the molecular mechanisms behind circadian oscillators, and the evolutionary significance of this remarkable adaptation.

But first, a natural question that may have crossed your mind; how was it discovered that plants had an internal clock? The next section traces the evolution of circadian rhythm research, from early observations to the modern era of molecular genetics and computational modeling.

1.1 *Evolution of the Study of Circadian Rhythms*

Cellular life originated approximately 3.7 billion years ago. With few exceptions, life on Earth has evolved under the consistent daily cycles and seasonal cycles resulting from the planet's rotation and revolution around the sun. The ability to anticipate these environmental changes has provided a significant advantage, leading to the development of internal circadian rhythms that align physiological processes with external conditions. However, unraveling the evolutionary history of these rhythms has been challenging, as the genes and proteins associated with biological clocks are not universally conserved across the three domains of life: bacteria, archaea, and Eukaryota [20]. The genetic networks that control the internal clock are mostly conserved across bacteria, plants, and humans, although the specific components within these networks vary. These circadian pacemakers function by roughly estimating time, requiring daily resetting to maintain synchronization between the internal and external clocks. Given this, it is unsurprising that light serves as the primary entrainment signal for circadian systems across a wide range of organisms, from single-celled life forms to mammals [7].

1.1.1 Early Observations: The First Evidences of an Internal Clock Keeping Mechanism

Long before we had microscopes or molecular biology, humans observed the daily rhythmic behavior of plants. One of the first known scientific references to circadian rhythms comes from *Androsthenes* (Fourth century BC), a Greek explorer who documented rhythmic leaf movements in tamarind trees, suggesting that plants had an internal sense of time [42].

However, controlled experiments demonstrating an endogenous clock came much later. Jean-Jacques d'Ortous de Mairan (1729) conducted a famous experiment on *Mimosa pudica*, a plant known for its folding leaf movements. He placed the plant in constant darkness and found that its leaves continued to open and close rhythmically, despite the absence of external light cues [44, 72]. This proved that the rhythm was internally generated rather than a direct reaction to light.

1.1.2 The 19th and 20th Centuries: Temperature Compensation and the Nature of Biological Clocks

In the nineteenth century, researchers discovered another defining feature of circadian rhythms: temperature compensation. Unlike most biological reactions, which accelerate at higher temperatures, circadian rhythms remain stable across different temperature ranges [15]. This ensures that plants maintain precise timing regardless of climate.

By the mid-twentieth century, three fundamental characteristics of circadian rhythms were established:

Periodicity—Circadian rhythms persist even in constant conditions, cycling approximately every 24 hours [16, 42].

Entrainment—The internal clock can be reset by external cues (light, temperature) [16, 42, 46].

Temperature Compensation—The period of the rhythm remains stable despite environmental fluctuations [16, 42].

However, it wasn't until the late twentieth century that scientists began to identify the genetic components of the plant circadian clock.

1.1.3 The Molecular Era: Genes Involved in the Circadian Clock

With advancements in molecular genetics, scientists identified the genes responsible for circadian regulation (Fig. 1):

Studies in *Neurospora crassa* (fungus) and *Drosophila melanogaster* (fruit flies) revealed the existence of transcription-translation feedback loops (TTFLs)—a core mechanism underlying circadian regulation [46].

By the 1990s and early 2000s, research on *Arabidopsis thaliana* identified key plant circadian clock genes, including:

TIMING OF CAB EXPRESSION 1 (TOC1)—A central regulator of the plant clock and *CIRCADIAN CLOCK ASSOCIATED 1 (CCA1) & LATE ELONGATED HYPOCOTYL (LHY)*—Morning-phase genes controlling rhythmic gene expression [9].

Overview: Timeline: Evolution of Circadian Rhythm Research in Plants.
The timeline begins with seed germination, marking the inception of both plant growth and early insights into rhythmic biological behavior. Just as the seed absorbs water, swells, and the radicle emerges, historical records—such as those by Androsthenes in the fourth century BCE—noted daily leaf movements in tamarind trees, hinting at an innate rhythmic mechanism. With seedling emergence, the shoot breaks through the soil and initiates photosynthesis. This parallels Jean-Jacques d'Ortous de Mairan's pivotal 1729 experiment with Mimosa pudica, where leaf movements persisted in constant darkness, revealing the presence of an endogenous timekeeping system.

As leaf development progresses and true leaves unfold, enhancing photosynthetic capacity, the scientific community in the nineteenth century began to recognize that circadian rhythms persist even under varying temperatures. This led to the identification of temperature compensation, a defining characteristic of circadian clocks. During root system expansion, when the plant anchors itself and increases nutrient absorption, mid-20th-century researchers established the formal concept of biological clocks. These were defined as

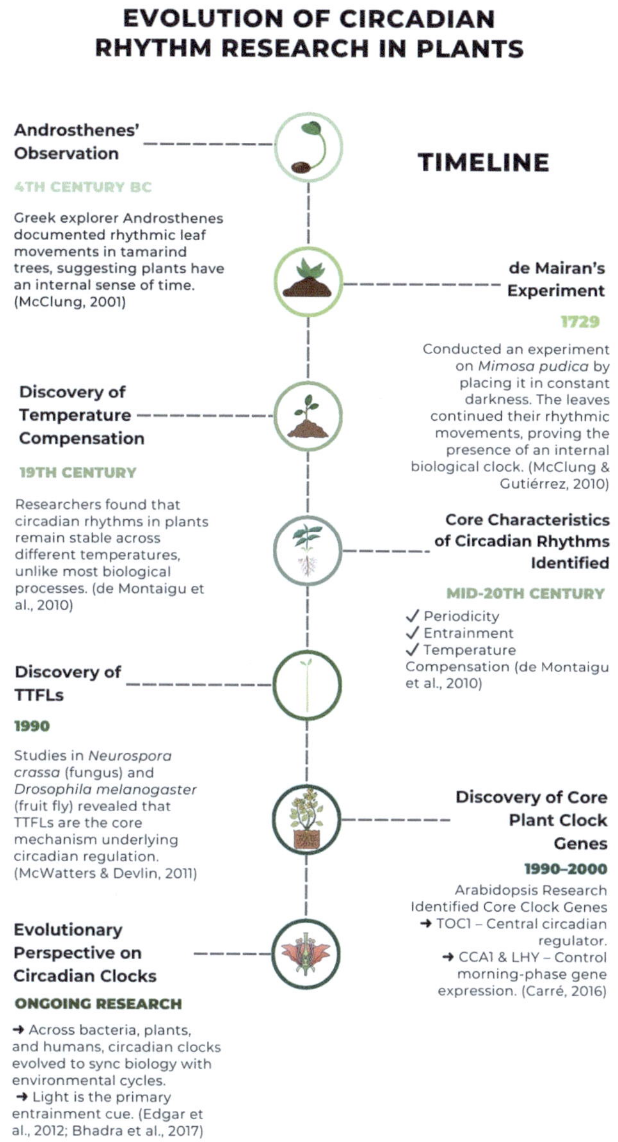

Fig. 1 Chronological timeline of plant circadian rhythm research depicted through the stages of plant development. This infographic illustrates a chronological progression of landmark discoveries in plant circadian rhythm research, visually embedded within the seven developmental stages of a plant's life cycle. The spine of the timeline is creatively represented by a growing plant, symbolizing the intrinsic connection between temporal regulation and biological development. Each phase in the plant's journey—from seed to seed—metaphorically reflects the deepening of scientific understanding in circadian biology

endogenous, self-sustained, entrainable oscillators, essential for synchronizing internal processes with external cycles.

With stem and shoot growth, the plant elongates and branches out, mirroring the expansion of molecular models explaining circadian function. In the 1990s and early 2000s, key clock genes such as CCA1, LHY, and TOC1 were identified in Arabidopsis thaliana. These discoveries revealed the architecture of transcription-translation feedback loops (TTFLs) and illuminated how internal clocks integrate environmental signals like light to regulate physiological processes.

The phase of flower formation denotes reproductive maturity, meticulously timed by the circadian clock to align with optimal environmental conditions. Research during this period uncovered how the clock regulates flowering through genes like GI and CO, linking timekeeping to photoperiodic responses. Finally, in pollination and seed production, the plant completes its life cycle, reflecting recent advances in the field. Contemporary findings underscore the universality of circadian systems across life forms and emphasize light as a fundamental entrainment signal. These discoveries bring the narrative full circle, reinforcing the idea that circadian clocks are a deeply conserved and indispensable feature of living systems.

1.2 A Generalized Scheme of Circadian Rhythms in Plants

Circadian rhythms are endogenous biological cycles that regulate physiological, metabolic, and behavioral processes over an approximately 24-h period. These rhythms persist even in the absence of external cues but can be entrained or synchronized by environmental signals such as light, temperature, and nutrient availability [43].

A defining feature of circadian rhythms is their ability to continue oscillating under constant conditions, a phenomenon known as free-running rhythm. This distinguishes them from diurnal rhythms, which require alternating light–dark cycles to persist [9]. Circadian rhythms are also temperature-compensated, meaning their period length remains relatively stable across a range of physiological temperatures, preventing fluctuations in timing due to seasonal changes [46]. The plant circadian rhythm is a highly coordinated system that regulates daily biological activities by synchronizing internal molecular processes with environmental cycles.

So how does this intricate timekeeping system work?

At its core, the plant circadian clock consists of three fundamental components—the core oscillator, input pathways, and output pathways. Together the three interconnected components: input pathways (which sense external signals), a central oscillator (which maintains a self-sustaining rhythm), and output pathways (which control physiological responses), ensure that plants anticipate and adapt to daily changes in light, temperature, and other environmental conditions.

1.2.1 The Input Pathways: Synchronizing the Clock with the Environment

Although the core oscillator maintains an endogenous rhythm, it must be synchronized (entrained) with real-world environmental changes to ensure accuracy and adaptability. Input pathways are essential for synchronizing the plant circadian clock with daily and seasonal environmental changes, ensuring that biological processes align with external conditions. These pathways allow plants to adjust their internal rhythms in response to key environmental cues such as light, temperature, and metabolic signals, maintaining accuracy and efficiency in physiological functions. By enabling adaptive plasticity, input pathways help plants thrive in fluctuating environments, optimizing energy use, growth, and reproduction. Additionally, they serve as critical regulators of stress responses, integrating information about energy status, nutrient availability, and external stressors into the circadian system. This ensures that plants can preemptively activate protective mechanisms against drought, cold, and pathogen attacks, ultimately enhancing survival and fitness.

The primary external cues influencing the circadian clock are:

Light: The Primary Zeitgeiber ("Time Giver")

Light is the strongest entrainment factor for plant circadian rhythms, allowing synchronization with the natural day-night cycle. A close relationship between light and the photoreceptors has been demonstrated in many systems [16]. Different light wavelengths are detected by specialized photoreceptors [69]. These photoreceptors detect and process ambient light signals, transmitting them to regulate plant growth and development. In plants, cryptochromes perform a diverse group of functions, including the entrainment of the circadian clock [11]. In Arabidopsis, the red/far-red light receptors phytochromes (PHYA and PHYB), have been shown to play a role in circadian entrainment [10, 66].

Phytochromes (PHYs) → Detect red (R) and far-red (FR) light, ensuring alignment with sunrise and sunset. [16, 19, 25, 41]

Cryptochromes (CRYs) → Sense blue light, adjusting the clock in response to changes in light intensity [16, 19, 41].

ZEITLUPE (ZTL) → A light-sensitive F-box protein that regulates the degradation of clock proteins, refining timekeeping [27, 61].

Mechanism of Light Entrainment:

During dawn, light-activated photoreceptors reset the core oscillator, increasing CCA1 and LHY expression to maintain morning-phase timing. As the day progresses, TOC1 accumulation is fine-tuned by light input, ensuring the transition to evening-phase gene expression.

Temperature: Another Timing Mechanism

Temperature fluctuations serve as a secondary entrainment cue, particularly for seasonal adjustments. Unlike light, which provides immediate feedback, temperature modulates the fine-tuning of circadian oscillations over longer timescales [15]. Warm temperatures accelerate the clock, while cool temperatures slow it down. Some PRR (PSEUDO-RESPONSE REGULATOR) proteins help mediate these responses, linking temperature signals to circadian regulation.

Key players in temperature regulation:

PRR7 and PRR9 → Act as temperature-sensitive regulators, ensuring stability in circadian rhythms despite seasonal variation.

Heat Shock Proteins (HSPs) → Protect the clock from temperature-induced disruptions.

Metabolic Signals and Hormones

Plants integrate metabolic status and hormonal signaling into their circadian system to adjust physiological responses. Sugar production from photosynthesis acts as a secondary clock regulator. During the day, sugars accumulate and feed back to influence clock gene expression [51]. This ensures that energy availability is aligned with biological processes like growth and flowering therefore sugar especially sucrose concentration also acts as an important zeitgeber in circadian rhythm regulation.

- Sugars and Carbon Metabolism:
  - Photosynthetically derived sugars provide feedback to the clock, influencing gene expression to balance energy demand and supply.
  - Sugar accumulation can modulate PRR7 expression, adjusting clock phase timing [46].
- Hormonal Crosstalk:
  - Abscisic Acid (ABA) → Governs drought responses, aligning stress-related processes with circadian oscillations [70].
  - Gibberellins (GA) → Control growth and flowering time, ensuring reproductive success in seasonal environments [46].
  - Cytokinins (CKs) → Utilize response regulators (RRs) in a prokaryotic-type phosphorelay system to mediate cell division, shoot initiation, light responses, and leaf senescence [47].
  - Pseudo-Response Regulators (PRRs) → Key components of the circadian clock, synchronizing biological rhythms with environmental light cues to regulate growth and flowering [47].
  - Histidine Kinases (HKs) & MAPKs → Bridge prokaryotic and eukaryotic signaling, coordinating responses to ethylene and other plant hormones [47].
  - Light-Signal Integration → PRRs function as internal "clocks," ensuring plants anticipate dawn and seasonal transitions (e.g., flowering before winter) [47].

1.2.2 The Core Oscillator: The Central Time Keeping Mechanism

The core oscillator is the heart of the circadian clock—a self-sustained molecular transcriptional-translational feedback loop (TTFL), that generates oscillations in gene expression and protein activity over an approximately 24-h cycle [6]. This intrinsic mechanism ensures that plants can maintain consistent daily rhythms even in the absence of external time cues [44]. The main components include:

- Morning-Phase Genes: Activation of CCA 1 and LHY

As dawn breaks, light activates CCA1 and LHY, two transcription factors that function as morning regulators. These proteins bind to DNA sequences in target genes and regulate their expression. A key function of CCA1 and LHY is to suppress TOC1 (TIMING OF CAB EXPRESSION 1), preventing its early activation [6, 13, 29, 54]

Midday Phase: PRR7 and PRR9 Modulate the Cycle

As the morning progresses, PRR7 and PRR9 genes start to be expressed. These genes help reduce CCA1 and LHY levels, allowing the next phase of the cycle to begin. By midday, CCA1 and LHY levels drop significantly, shifting control to other genes.

- Evening-Phase Genes: TOC 1 and Night Time Activation

As daylight fades, TOC1 levels start to rise, since its suppression by CCA1 and LHY is weakened. TOC1 acts as an evening regulator, preparing the plant for night time physiological changes [55]. TOC1 and other evening genes eventually contribute to reactivating CCA1 and LHY by morning, ensuring the cycle repeats [6, 9]. This continuous sequence of gene activation and repression ensures that the clock maintains a 24-h rhythm, independent of environmental changes.

- Regulatory Proteins and Additional Loops:

PSEUDO-RESPONSE REGULATORS (PRR5, PRR7, PRR9) act as negative regulators, sequentially repressing CCA1 and LHY throughout the day to refine rhythm timing.

The Evening Complex (EC), consisting of EARLY FLOWERING 3 (ELF3), ELF4, and LUX ARRHYTHMO (LUX), regulates TOC1 suppression at night, maintaining precise oscillations [15] (Table 1).

1.2.3 Output Pathway: Orchestrating Plant Responses

Once the core oscillator is synchronized through input pathways, the circadian system regulates multiple downstream processes, aligning them with optimal times of day. Output pathways play a crucial role in translating circadian rhythms into tangible physiological and developmental responses, optimizing resource allocation and overall plant fitness. By regulating processes such as photosynthesis, starch metabolism, and energy storage, the circadian clock ensures that plants use

Table 1 The roles of gene involved in the circadian clock of plants

Gene	Role in clock	Most active time	Effect on other genes
CCA1 (circadian clock associated 1)	Helps start the clock in the morning by activating daytime genes and blocking evening genes	Early morning	Turns on PRR7, PRR9 and suppresses TOC1
LHY (late elongated hypocotyl)	Works alongside CCA1 to regulate the clocks rhythm	Early morning	Turns on PRR7, PRR9 and suppresses TOC1
PRR7 (pseudo-response regulator 7)	Contributes to reducing CCA1 and LHY levels as the day progresses	Midday	Slows down CCA1 and LHY production
PRR9 (pseudo-response regulator 9)	Supports PRR7 in adjusting CCA1 and LHY levels	Midday	Suppress CCA1, LHY expression
TOC1 (timing of CAB expression 1)	Builds up in the evening and helps restart the cycle for the next day	Evening/ night	Increases when CCA1/ LHY are low and helps start their production

resources efficiently, maximizing growth and minimizing energy waste [42]. Additionally, circadian-controlled flowering pathways synchronize reproductive events with seasonal photoperiods and pollinator availability, increasing the chances of successful fertilization [53, 68]. Beyond growth and reproduction, output pathways also enhance stress resilience by anticipating environmental challenges such as drought, extreme temperatures, and pathogen attacks. Through precise timing of stress response genes, the circadian clock allows plants to activate protective mechanisms before adverse conditions occur, improving survival and adaptability in dynamic ecosystems [28].

Photosynthesis and Energy Metabolism

The circadian clock maximizes photosynthetic efficiency by regulating:

- Stomatal opening → Ensures optimal CO_2 uptake in response to light availability.
- Chlorophyll biosynthesis → Peaks in the morning when light capture is highest.
- Starch metabolism → Prevents premature depletion of carbohydrate reserves overnight [9].

Growth and Development

The timing of cell division, elongation, and differentiation is tightly controlled by circadian rhythms.

- Hypocotyl Elongation:

PIF4 and PIF5 (growth regulators) are circadian-controlled, promoting stem elongation before dawn when light conditions are favorable [46].

- Leaf Movement and Nyctinasty:

Some plants adjust leaf orientation rhythmically, optimizing light capture during the day and minimizing water loss at night [48, 62].

Flowering Time

Circadian clocks regulate photoperiodic flowering, ensuring reproduction occurs in favorable seasonal conditions.

- CONSTANS (CO) & FLOWERING LOCUS T (FT) Pathway:
  - In long-day plants, CO accumulates in response to extended daylight, triggering FT expression and initiating flowering.
  - The clock ensures CO accumulation aligns with light availability, preventing premature flowering in winter [44].

Stress Responses

The circadian clock anticipates environmental stressors, regulating:

Cold resistance genes (CBF pathway) → Activated before nightfall, preparing plants for freezing temperatures [50]

Pathogen defense responses → Many immune-related genes peak in expression at night, when infection risk is highest [15].

Together, these components form a highly dynamic system, one that is not only vital for individual plant survival but also shapes entire ecosystems, agricultural yields, and even global carbon cycles. In the sections that follow, we will explore how these interconnected pathways allow plants to stay in tune with the planet's ever-changing rhythms, ensuring that they thrive in an unpredictable world (Fig. 2).

1.3 Understanding the Circadian Rhythms in the Model Plant Arabidopsis Thaliana

Arabidopsis thaliana, commonly known as Thale cress, is a member of the mustard family, typically found in parts of Europe, Asia, and Africa. Over the years, it has become the premier model organism for circadian rhythm research due to its several advantageous traits. These include its small size, rapid seed production, a short generational cycle of about 6–8 weeks, and a compact genome. These characteristics make it particularly convenient for researchers aiming to understand the complexities of circadian biology. *Arabidopsis* is a monoecious plant, meaning it can self-fertilize, though outcrossing is also possible [57].

A properly functioning circadian clock bestows several advantages on *Arabidopsis* that significantly enhance its overall fitness and survival. One key benefit is improved survival rates under short-day conditions. Plants with a functional circadian clock thrive in such conditions, while mutants with disrupted rhythms, such as CCA1-ox

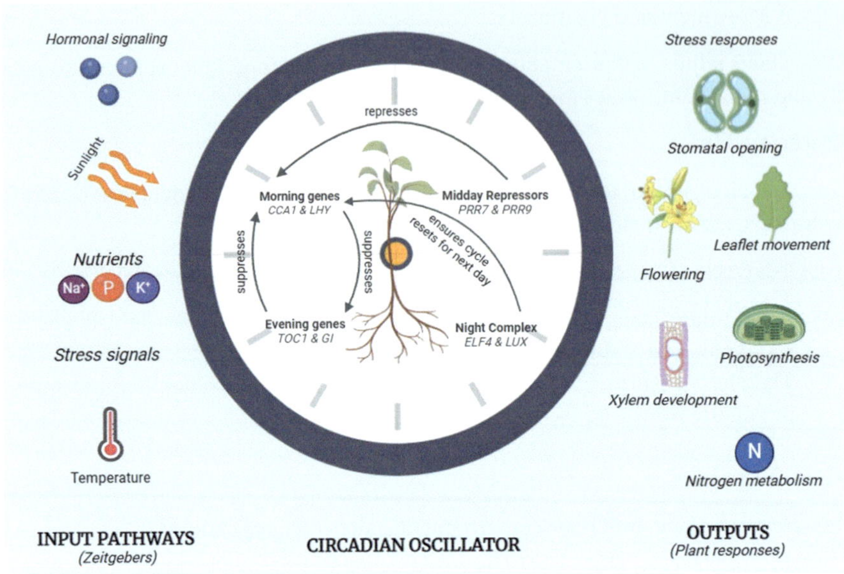

Fig. 2 An overview of the mechanism of circadian control in plants. The plant circadian oscillator is depicted as a clock, illustrating the transcriptional feedback loops that regulate rhythmic gene expression over a 24-h cycle. The central oscillator consists of morning genes (CCA1 & LHY), which suppress evening genes (TOC1 & GI), and are subsequently repressed by them to maintain cyclic regulation. Midday repressors (PRR7 & PRR9) and the night complex (ELF4 & LUX) contribute to sustaining rhythmic oscillations and phase transitions. Input pathways (Zeitgebers), shown on the left, include sunlight, temperature, hormonal signaling, nutrients (Na^+, P, K^+), and stress signals, all of which entrain the circadian clock by modulating gene expression. Output responses, illustrated on the right, represent physiological and developmental processes regulated by the circadian oscillator. These include stress responses, stomatal opening, leaflet movement, flowering, photosynthesis, xylem development, and nitrogen metabolism, all of which are optimized through circadian control to enhance plant growth and adaptation to environmental condition

mutants, struggle to grow. This suggests that synchronization with environmental cycles, particularly light and dark, enables plants to use energy more efficiently [60].

Another advantage is the regulation of flowering time. Wild-type *Arabidopsis* with a functioning circadian clock flowers at the appropriate time under long-day conditions. In contrast, mutants with altered circadian rhythms experience delayed flowering. The proper timing of flowering is critical for reproductive success, ensuring that pollination occurs when environmental conditions are most favorable [14].

Circadian regulation also allows plants to anticipate daily environmental changes. Wild-type *Arabidopsis* adjusts key physiological processes, such as photosynthesis, in advance of sunrise, optimizing resource use. Mutants lacking this anticipatory ability fail to synchronize effectively with light cycles, which could reduce their growth efficiency [33, 34].

1.3.1 Discovery of the Circadian Clock Mutants

The journey into understanding *Arabidopsis*' circadian rhythm began around 1996, with the discovery of the first circadian clock mutant. This discovery led to the identification of a remarkable number of 20 different genes involved in regulating the plant's circadian clock. This number was far larger than that observed in animals, where fewer genes were known to control the circadian rhythm. The reason for this abundance of genes lies in *gene redundancy*, a phenomenon where multiple genes serve similar functions, often with overlapping roles [53].

For instance, *Arabidopsis* contains five phytochromes—PHYA, PHYB, PHYC, PHYD, and PHYE—and two cryptochromes, CRY1 and CRY2, all of which play key roles in regulating the plant's circadian rhythm. While these proteins have overlapping functions, they are not entirely redundant. For example, PHY A is involved in the plant's response to light, especially far-red light, and modulates reactions to low levels of red and blue light [25]. Phytochromes PHY B, D, and E, on the other hand, are involved in lengthening the circadian period under constant red light, which mimics the effect of reduced light intensity. Meanwhile, cryptochromes CRY1 and CRY2 are crucial for regulating the plant's self-sustained circadian rhythm under constant blue light [19, 41]. Together, this network of genes and their overlapping functions equips *Arabidopsis* with a highly adaptable and flexible circadian clock, vital for its growth and survival in an ever-changing environment [27, 45].

1.3.2 The Role of ELF4 in Circadian Rhythm Regulation

Research by [19] highlighted the significance of the *EARLY FLOWERING 4* (ELF4) gene in *Arabidopsis*. ELF4 plays a critical role in maintaining the stability of the plant's internal biological clock and its flowering response. Their findings showed that ELF4 helps stabilize the circadian system even when external light–dark cycles are absent. Mutations in ELF4 led to early flowering under conditions that would typically delay it, likely due to an increase in CONSTANS (CO), a key promoter of flowering. Furthermore, disruption of ELF4 resulted in reduced expression of *CIRCADIAN CLOCK ASSOCIATED 1* (CCA1), a vital component of the circadian system, leading to a loss of regular rhythmic patterns. These findings underscore ELF4's pivotal role in regulating both circadian function and photoperiodic flowering in *Arabidopsis* [12, 19].

1.3.3 F-Box Proteins and Their Influence on the Circadian Clock

A crucial component of the circadian clock in *Arabidopsis* is the role of F-box proteins, which are involved in the targeted degradation of proteins through the ubiquitin–proteasome system. These proteins are part of the SCF (SKP1-CUL1-F-box) complex and help regulate the circadian cycle by selectively marking specific clock-associated proteins for degradation. This process ensures that key clock regulators are removed at precise intervals, maintaining the stability and timing of the circadian cycle [27].

F-box proteins are not only essential for circadian regulation but also for responding to external environmental cues, such as changes in light and temperature [24, 58]. By adjusting the circadian clock in response to these factors, plants can fine-tune their internal biological cycles to better align with seasonal variations. Additionally, F-box proteins are involved in broader regulatory networks, influencing processes such as flowering time, leaf movement, and stress responses. Disruptions in F-box protein function can lead to unstable circadian rhythms, negatively impacting plant growth, photosynthesis, and hormone signaling [49, 63] (Fig. 3).

Morning—Wake-Up Genes: At dawn, the transcription factors CCA1 and LHY are expressed, initiating the day by triggering stomatal opening. This facilitates gaseous exchange and primes the plant for photosynthesis. Afternoon—Time to Work Genes: As the day progresses, genes such as PRR9, PRR7, and PRR5 reach peak expression. These regulators modulate key processes like flowering time and energy metabolism. Evening—Relax and Prepare Genes: During twilight hours, the evening complex—composed of ELF3, ELF4, and LUX—becomes active. These genes facilitate night time preparation by promoting cell elongation and growth, symbolized as "Reaching for the Sun." Night—Time to Rest Genes: In the dark phase,

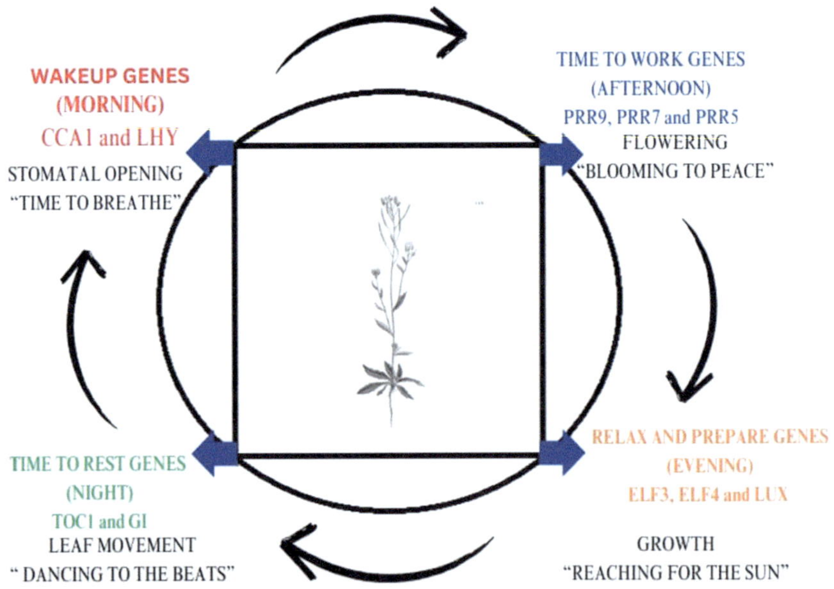

Fig. 3 This schematic illustrates the temporal expression patterns of core circadian clock genes in *Arabidopsis thaliana* across a 24-h cycle. As a model organism in chronobiology, *A. thaliana* displays a highly regulated internal clock that aligns physiological and molecular activities with external light–dark cues. The diagram is organized into four primary temporal zones

genes such as TOC1 and GI govern rhythmic leaf movement and support restorative functions, reflected in the expression "Dancing to the Beats." The cyclical layout emphasizes the interconnected feedback loops that sustain rhythmic gene expression. These oscillations optimize vital functions like photosynthesis, growth, and reproduction, enhancing the plant's adaptability to its environment.

1.4 Key Benefits of Circadian Rhythm Regulation in Arabidopsis

The circadian rhythm system offers numerous advantages to plants like *Arabidopsis*:

Environmental Synchronization: The circadian clock aligns the plant's internal biological processes with external environmental cycles, ensuring that physiological responses are optimized for daily changes in light and temperature [37, 67].

Photosynthesis Regulation: By timing light-harvesting and carbon assimilation processes, the circadian clock maximizes photosynthetic efficiency, ensuring that the plant uses light and energy at the most favorable times of the day [39, 41].

Metabolic Coordination: The circadian clock regulates essential metabolic pathways, including sugar metabolism, starch breakdown, and energy distribution, all of which contribute to optimal growth and resource allocation [53].

Hormonal Control: The timing of hormone biosynthesis and signaling is controlled by the circadian rhythm, affecting plant development, stress responses, and overall physiological processes, including auxins, gibberellins, and abscisic acid [45].

Growth and Development: Key developmental processes, such as hypocotyl elongation, leaf movement, and root growth, are governed by the circadian clock, helping the plant respond adaptively to environmental cues [26, 58].

Gene Expression Regulation: The circadian system regulates the expression of thousands of genes, ensuring that biological pathways are activated at the correct times for maximum efficiency [12, 59, 64].

Stress Response Optimization: Circadian rhythms enhance the plant's ability to resist abiotic stress (such as drought and temperature extremes) and biotic stress (such as pathogen attacks), thereby improving overall resilience [1, 14]

Flowering Time Control: By regulating genes such as *CONSTANS* (CO) and *FLOWERING LOCUS T* (FT), the circadian clock controls the timing of flowering, ensuring that reproductive processes occur in optimal environmental conditions [32, 40, 42].

Protein Turnover and Stability: Circadian regulation plays a vital role in maintaining protein turnover and stability, particularly through mechanisms like ubiquitin-mediated proteolysis, ensuring that the plant's circadian system remains precise and stable [24, 67].

Agricultural and Biotechnological Applications: Understanding circadian rhythm regulation in *Arabidopsis* has far-reaching applications in agriculture and biotechnology. Insights into circadian control can inform crop breeding efforts aimed at enhancing yield, stress tolerance, and adaptability to changing environmental conditions [12, 39].

Arabidopsis thaliana, has proven to be an invaluable model for understanding circadian rhythms in plants. Through a combination of gene redundancy, intricate regulatory networks, and external cues, *Arabidopsis* exemplifies how circadian regulation is essential for plant growth, development, and survival. The insights gained from this model organism have far-reaching implications for improving agricultural productivity and resilience in the face of environmental challenges. By delving deeper into the molecular mechanisms underlying circadian regulation, researchers continue to unlock new strategies for enhancing crop performance, making *Arabidopsis* a key player in the future of sustainable agriculture.

1.5 Timed to Thrive: Rhythmic Behavior Exhibited by Different Plant Processes

A number of plant processes exhibit rhythmic behavior. The daily cycle of light and darkness profoundly impacts physiological processes in plants.

1.5.1 Oscillations in Chlorophyll Content

Chlorophyll content in plants exhibits temporal fluctuations influenced by various environmental and physiological factors. These oscillations arise due to diurnal cycles, seasonal variations, and intrinsic biological rhythms. Recent research highlights the mechanisms governing these fluctuations, emphasizing the interplay between light availability, photosynthetic efficiency, and metabolic regulation [5].

Chlorophyll content follows a diurnal cycle, typically peaking in the early morning and midday before declining in the evening [5]. These oscillations align with the plant's circadian rhythm, which modulates chlorophyll biosynthesis and degradation in response to light cues. The synthesis of chlorophyll precursors, such as protochlorophyllide, is synchronized with day light availability, ensuring optimal photosynthetic activity. Additionally, internal clocks regulate the expression of genes associated with chlorophyll synthesis, reinforcing the rhythmicity of pigment accumulation [5].

External environmental factors significantly impact chlorophyll fluctuations. Light intensity plays a crucial role, as higher exposure promotes chlorophyll accumulation, while shaded conditions reduce synthesis [5]. Temperature variations also influence enzymatic activities related to chlorophyll metabolism—higher temperatures accelerate degradation, whereas moderate temperatures support stability and synthesis. Furthermore, seasonal changes dictate chlorophyll content, with peaks observed during periods of active growth, such as spring and summer, and declines during autumn as plants transition into dormancy [5].

Oscillations in chlorophyll content are closely tied to photosynthetic efficiency. During periods of high photosynthetic demand, chlorophyll levels rise to maximize energy capture. Conversely, under stress conditions such as drought or nutrient deficiency, degradation mechanisms activate to conserve resources. Additionally, metabolic pathways involving nitrogen and magnesium—essential components of chlorophyll molecules—exhibit rhythmic fluctuations that correspond with chlorophyll oscillations. These processes are further modulated by plant hormones like cytokinins and gibberellins, which influence chlorophyll biosynthesis and stability [5].

1.5.2 Xylem Development

Xylem development in trees is tightly regulated by environmental and physiological factors, with diurnal and circadian rhythms playing a crucial role in modulating gene expression and metabolic pathways. The daily cycle of light and darkness profoundly impacts physiological processes in plants, including carbon allocation, hormone signaling, and wood formation [56].

The circadian clock governs gene expression in plants, enabling them to anticipate daily environmental changes and optimize physiological processes accordingly [56]. Studies on *Eucalyptus* trees have identified diurnal expression patterns in wood-forming tissues, with approximately 8% of genes exhibiting rhythmic fluctuations in transcript abundance. These genes are involved in carbon allocation, stress responses, and secondary metabolism, highlighting the complex interplay between light availability and xylem development [56]. The presence of central clock genes, such as *Circadian Clock Associated 1 (CCA1)* and *GIGANTEA (GI)*, in developing xylem suggests that temporal control mechanisms are integral to wood formation and metabolic regulation [56].

Both primary and secondary metabolic pathways in plants are influenced by diurnal cycles. The availability of light directly impacts photosynthesis and carbon fixation, thereby affecting the coordination of carbon supply and utilization during xylem development. Additionally, lignin biosynthesis—a key process in xylem differentiation—is temporally regulated, with phenylpropanoid metabolism exhibiting rhythmic expression patterns [56]. Such cyclic regulation ensures efficient resource allocation and structural integrity of developing xylem tissues.

Beyond daily fluctuations, the circadian system also influences long-term developmental processes such as seasonal growth, dormancy, and flowering. These prolonged changes are essential for the survival of woody perennials, as they allow trees to adapt to environmental variations over extended periods. Several genes involved in xylem differentiation, cell wall biosynthesis, and carbohydrate metabolism exhibit rhythmic expression. The homologs of core circadian clock genes (*EtCCA1, EgGI, EtZTL*) show diurnal fluctuations, suggesting that the molecular clock regulates secondary growth in trees [56]. The peak expression of genes at specific Zeitgeber Times (e.g., ZT 16, 22:00) indicates synchronized metabolic activity that optimizes growth efficiency.

Genes related to cellulose synthesis, lignin biosynthesis, and carbohydrate metabolism follow distinct rhythmic patterns. Carbohydrate-active enzymes (CAZymes), crucial for sugar metabolism, show diurnal peaks, aligning with carbon availability [56]. Lignin biosynthesis genes peak before dawn and mid-afternoon, suggesting a regulated pattern for reinforcing secondary cell walls.

The central circadian clock in *Eucalyptus* wood-forming tissues aligns with *Arabidopsis*, suggesting conservation across herbaceous and woody species [56]. Key genes like *CCA1* and *GI* show similar expression patterns in *Eucalyptus* and *Arabidopsis*. Unlike *Arabidopsis*, where *ZTL* is regulated at the protein level, *Eucalyptus ZTL* is controlled at the transcriptional level.

Plants balance carbon levels between day (surplus) and night (deficit), with large trees facing additional challenges due to spatial separation of carbon sources and sinks. Genes related to cellulose, hemicellulose, and lignin biosynthesis show diurnal variation in *Eucalyptus*, aligning with *Arabidopsis* and suggesting circadian control over cell wall formation processes [56].

The diurnal expression patterns of genes involved in carbon allocation, cell wall biosynthesis, and lignin formation suggest that the internal biological clock plays a crucial role in coordinating wood formation with environmental cues.

1.5.3 Root Growth

While the role of circadian rhythms in leaf movement, photosynthesis, and flowering is well-documented, their impact on root growth remains an emerging area of study. Recent findings indicate that root elongation follows a 23-h circadian cycle along with 2-h ultradian oscillations, demonstrating that root development is not continuous but regulated by rhythmic biological clocks [31].

Circadian rhythms in root growth enable plants to synchronize their developmental processes with external light–dark cycles. Studies show that root elongation rates fluctuate based on light exposure and temperature variations [31].

These variations suggest that daytime root growth benefits from enhanced transpiration and carbon supply, while night time growth slows to conserve energy. Light perception pathways, mediated by phytochromes (PHYs) and cryptochromes (CRYs), regulate auxin transport, suppressing root elongation at night. This ensures a coordinated balance between shoot and root growth in response to environmental signals.

Roots depend on carbon supplied from photosynthesis, which follows a circadian pattern. During the day, carbohydrate availability increases due to active **sucrose transport via the phloem**, supporting maximum root elongation. Key molecular regulators include:

SUCROSE TRANSPORTER1 (SUC1): Expressed rhythmically to enhance carbohydrate movement to root tips [31].

STARCH EXCESS1 (SEX1) and GRANULE-BOUND STARCH SYNTHASE (GBSS1): Control starch breakdown at night, ensuring a steady energy supply [31].

Plant growth regulators like auxins and ethylene also have an impact on root elongation. Both these hormones are under the circadian control. Auxin, a central hormone in root elongation, exhibits circadian oscillations that regulate its production and distribution. The genes that play an important role in auxin cycles are: TRYPTOPHAN AMINOTRASNFERASE OF ARABIDOPSIS (TAA 1), PIN-FORMED (PIN1, PIN2) and AUX1. Circadian-controlled auxin transporters that facilitate polar auxin flow, ensuring rhythmic growth cycles [31].

Ethylene, known for its inhibitory effects on root elongation, also follows circadian patterns. Beyond the 23-h circadian cycle, root elongation exhibits a 2-h ultradian rhythm, possibly linked to rhythmic secretion of mucilage and border cell release. These short-term oscillations have been observed across various species like climbing beans, sunflower, cucumber, radish, and maize, etc. [31].

1.5.4 L-ascorbic Acid Levels

L-ascorbic acid (Asc), commonly known as vitamin C, plays a crucial role in plant physiology, particularly in stress response, growth, and photosynthesis. *Euglena*, a unicellular photosynthetic organism, exhibits circadian rhythms in Asc levels, independent of cell cycle progression [36]. The key findings related to circadian regulation of Asc in Euglena include:

Peak Synthesis at Subjective Dawn: Maximum Asc synthesis occurs at subjective dawn, influencing overall Asc levels [36]. In higher plants, such as spinach, circadian rhythms in Asc content have been observed, particularly in leaf tissues [36]. The rhythmic fluctuation of Asc levels suggests an adaptive mechanism for protecting plants from oxidative stress. The peak in Asc levels at midday coincides with the highest photosynthetic activity and ROS production, providing antioxidant protection [36]. UV Resistance: Although Asc contributes to UV protection, the circadian rhythm of UV resistance persists even in darkness, implying additional regulatory factors beyond Asc rhythms. While photosynthesis influences Asc synthesis, its rhythmic pattern is governed by an intrinsic biological clock [36]. The synchronization of Asc levels with midday oxidative stress suggests an evolutionary adaptation for enhanced protection.

1.5.5 Pollen Emissions

Pollen emission in corn exhibits a strong diurnal pattern, with peak release occurring in the morning and declining in the afternoon [64]. This pattern aligns with solar radiation levels, which drive anther desiccation, facilitating pollen dispersal. The first peak in pollen concentration was observed just before maximum irradiance on the anthers, suggesting that sunlight is a key trigger for pollen release.

Temperature and humidity play crucial roles in pollen emission. High humidity and low temperature suppress pollen release by preventing anthers from drying, whereas decreasing humidity and increasing temperature promote anther desiccation and pollen dispersal [64]. The concentration of pollen release in the morning provides several evolutionary benefits for maize including: avoidance of night time moisture, increased pollination success and protection from environmental stress. Angevin et al. [2] estimated seasonal pollen emission fluxes by integrating an empirical seasonal pattern specific to each plant species with the distribution of emission onset dates for individual plants. Diurnal patterns of Poaceae and Betula pollen flight have recently been reported by Marceau [38].

1.5.6 Leaflet Movement

The adaptive significance of circadian leaf movements lies in optimizing a plant's sunlight exposure during the day by raising its leaves while reducing water loss and protecting against cold night temperatures by lowering them at night. This adaptation aligns with the daily light–dark cycle, enhancing photosynthetic efficiency. These movements are governed by an internal clock that responds to environmental cues such as light and temperature. Research demonstrates that *Marama* bean leaflets exhibit a regular rhythmic movement pattern under both natural light–dark cycles and continuous darkness [62]. This persistence under constant conditions suggests the presence of an internal biological clock that drives leaflet motion. The leaflet movements follow a clear diurnal pattern, with opening during the day and closure at night, aligning with a typical circadian cycle [62]. It was observed that under normal conditions, leaflet angles increase during daylight hours and decrease at night, showing a consistent oscillatory pattern over multiple days. In constant darkness, rhythmic leaflet movement continues, confirming that the mechanism is endogenously controlled rather than being solely a response to external light stimuli [62]. [48] reported that the daily periodic leaf movements are striking examples of "osmotic motors" controlled by the endogenous biological clock and influenced by light, which is detected by phytochrome and possibly phototropins. Circadian control of leaf movements have been reported in several plants, including the model plant system of *Arabidopsis* [21].

Not just light, but temperature also acts as a zeitgeber in regulating leaflet movements. Under varying temperature conditions (ranging from low to high temperatures), rhythmic movement persists, although minor fluctuations are observed. The consistency of leaflet motion across different temperature regimes further supports the hypothesis that circadian rhythms regulate these movements, as external temperature changes do not completely override the rhythmic pattern [62].

1.5.7 Flowering and Photoperiodic Control

Flowering in plants is a highly regulated process influenced by both environmental and endogenous signals. Among these, photoperiodism and circadian rhythms play a crucial role in determining the appropriate time for floral initiation. Circadian rhythms control the expression of essential genes that drive the transition from vegetative growth to flowering. Key genes involved include CONSTANS (CO)—A gene that promotes flowering under long-day conditions and FOWERING LOCUS T (FT), which acts downstream of CO and transmits the flowering signal to the shoot apical meristem [23].

Plants measure day length using their circadian clocks to distinguish between long-day and short-day conditions, ensuring that flowering occurs at the optimal time. This adaptation is crucial for maximizing reproductive success in varying seasonal environments [68]. A fine-tuned balance between flowering promoters and repressors prevents premature flowering.

The circadian clock influences histone modifications, leading to changes in gene expression patterns that regulate flowering time. These epigenetic mechanisms allow plants to adjust their flowering schedules in response to environmental changes [23]. Circadian rhythms help plants integrate multiple environmental cues, such as: temperature: Modulates flowering time in conjunction with photoperiodic cues, light intensity and quality: Influences the activation of flowering-related genes and seasonal changes: Enable dynamic adjustments in flowering schedules [72]. Seasonal changes in day length are perceived by photoreceptors and transmitted to the circadian clock to regulate flowering time. Key genes involved in this process include, COP1 (CONSTITUTIVE PHOTOMORPHOGENIC 1), ELF3 (EARLY FLOWERING 3) and GI (GIGANTEA. ELF3 enables COP1 to interact with GI, leading to its degradation and affecting light input to the circadian clock [71].

2 Comparative Study: Plant Versus Animal Circadian Rhythms

The Hidden Timekeepers of Life

Life on Earth operates in synchrony with the 24-h cycle of day and night. Organisms across all domains have evolved biological clocks—internal timekeeping systems that regulate their behaviors, metabolism, and physiology. From the blooming of flowers to the sleep–wake cycles of mammals, circadian rhythms help organisms anticipate and adapt to environmental changes.

Despite their shared reliance on circadian rhythms, plants and animals have developed distinct mechanisms to regulate their internal clocks. Plants, being sessile, rely heavily on light as the primary environmental cue, orchestrating their photosynthesis, metabolism, and growth accordingly [42]. Animals, on the other hand, integrate multiple cues such as light, temperature, feeding cycles, and social interactions via a centralized circadian pacemaker located in the suprachiasmatic nucleus (SCN) of the hypothalamus [18].

Recent research has also revealed the role of cryptochromes, flavoproteins found in both plants and animals, in regulating circadian rhythms. In plants, cryptochromes act as blue light photoreceptors that mediate photoperiodic flowering and growth [10], while in mammals, they function as key components of the molecular clock, influencing sleep, metabolism, and behavioral rhythms [34].

Understanding the differences between plant and animal circadian rhythms is essential not only for advancing agriculture and crop production but also for addressing human health concerns related to circadian misalignment, such as sleep disorders, metabolic diseases, and mental health conditions [3].

This study will explore the molecular mechanisms, environmental adaptations, and physiological roles of circadian rhythms in plants and animals, offering insights into the universal principles of biological timekeeping Table 2.

3 Conclusion

Circadian rhythms are one of the most amazing and intriguing adaptations of life on earth. They are internal biological cycles that follow an approximately 24-h period, allowing plants to anticipate and adapt to daily environmental changes. These rhythms are governed by an endogenous clock that integrates external cues such as light, temperature, and humidity to regulate various physiological and developmental processes. In preparation for dawn, plants accumulate key components of the photosynthetic machinery daily. Meanwhile, leaves or flowers frequently close before nightfall, shielding delicate tissues from cooler night time temperatures.

The circadian clock does not function independently of the natural day-night cycle; it requires synchronization to maintain accurate timing. To achieve this, the clock incorporates a resetting mechanism that aligns it with environmental light–dark transitions, enabling the organism to anticipate dawn and dusk correctly. The entire plant machinery participates in this circadian clock. There are inputs to the resetting of the clock which are termed as zeitgebers, light and temperature being fundamental in nature. Upon receipt of these inputs plants entrain their internal clock which results in changes in their physiological, morphological, and behavioral outputs. For the successful functioning of these rhythms the genetic machinery of the plants participates in the end response.

Some of the key responses in plants which exhibit rhythmic behavior include.

- Leaf Movements—Many plants exhibit rhythmic leaf movements, raising their leaves during the day to maximize sunlight exposure and lowering them at night to conserve water and reduce heat loss.
- Photosynthesis and Stomatal Regulation—Circadian rhythms optimize photosynthesis by coordinating stomatal opening and closing, ensuring efficient CO_2 uptake while minimizing water loss, particularly during drought conditions.

Table 2 Comparative study of plant and animal circadian rhythms

Aspect	Plant circadian rhythm	Animal circadian rhythm
Molecular regulation	Plants utilize clock genes like TOC1, CCA1, and LHY to regulate their circadian cycles. These genes are involved in processes like photosynthesis and flowering	Animals rely on genes like CLOCK, BMAL1, PER, and CRY which form feedback loops controlling metabolic and behavioral rhythms [65]
Core genes	Key clock genes in plants include TOC1, CCA1, and LHY. These genes help regulate light sensing, photosynthesis, and timing of flowering	Core clock genes in animals are CLOCK, BMAL1, PER, and CRY, which regulate sleep–wake cycles, metabolic processes, and hormone release [17]
Light sensitivity	Plants are highly sensitive to light and use light as a primary cue for entrainment, affecting processes like photosynthesis and flowering [65]	In animals, light exposure influences the SCN (suprachiasmatic nucleus) in the hypothalamus, which regulates sleep and metabolism. Light exposure resets the biological clock [3]
Rhythm function	Plant rhythms optimize energy production (photosynthesis) and energy allocation for growth, with peak metabolic activity during daylight hours [65]	In animals, circadian rhythms regulate behavioral functions like feeding, sleep, hormone secretion, and metabolism. The central clock in the SCN synchronizes peripheral clocks [18]
Phase control	Plant rhythms optimize energy production (photosynthesis) and energy allocation for growth, with peak metabolic activity during daylight hours [65]	In animals, circadian rhythms regulate behavioral functions like feeding, sleep, hormone secretion, and metabolism. The central clock in the SCN synchronizes peripheral clocks [18]
Feeding and metabolism	In plants, circadian rhythms regulate the timing of photosynthesis, ensuring efficient light energy capture during the day. Starch storage and breakdown follow a rhythmic pattern [65]	In animals, circadian rhythms control feeding times and influence insulin sensitivity, with peak metabolic efficiency during the active phase of the day [18]
Hormonal regulation	In plants, hormonal processes like gibberellins and auxins are influenced by circadian rhythms, regulating growth and flowering time [65]	Animals exhibit hormonal regulation via the circadian control of cortisol, melatonin, and growth hormones, which are linked to sleep, stress, and metabolism [18]
Temperature regulation	Plant metabolic processes, such as enzyme activity, are regulated by circadian rhythms, optimizing energy use based on temperature variations during the day [65]	In animals, body temperature follows a daily rhythm, with body temperature typically lower during sleep and highest during the daytime [3]

(continued)

Table 2 (continued)

Aspect	Plant circadian rhythm	Animal circadian rhythm
Immune function	Plant circadian rhythms influence the timing of immune responses to pathogens, optimizing defense mechanisms during specific phases of the day [65]	Animal circadian rhythms modulate immune functions, with peak immune responses often occurring during the night, helping to fight off infections [18]
Adaptation to light	Plants can adapt their circadian rhythms to fluctuating light conditions, allowing them to optimize photosynthesis and growth. They also adjust to changes in the day length	In animals, light exposure, particularly in the morning, helps synchronize internal clocks, while light exposure in the evening can delay circadian rhythms, causing sleep disorders [3]
Environmental entrainment	Plants rely heavily on external light cues (zeitgebers) for synchronization, which directly influences their growth and flowering cycles	Animals also depend on environmental cues such as light, temperature, and social interaction to entrain their biological clocks, synchronizing behavior and metabolism [3]
Reproductive cycles	Plant circadian rhythms synchronize flowering, pollination, and seed dispersal with optimal environmental conditions, especially during specific photoperiods [65]	Animal circadian rhythms also control reproductive cycles, timing mating and birthing for seasons that offer better survival chances for offspring [65]
Enzyme activity	In plants, circadian rhythms regulate enzymes responsible for key metabolic functions like photosynthesis and starch breakdown, ensuring they peak during the day	In animals, enzyme activities related to metabolic processes, including glucose metabolism and fat oxidation, follow circadian rhythms, affecting digestion and nutrient processing [18]
Free-running period	Plants typically exhibit a free-running circadian rhythm of about 24 h, although environmental factors can cause slight variations	Animals, like humans, exhibit a free-running circadian period slightly longer than 24 h (around 24.2 h), typically aligning with the day/night cycle in a controlled environment [3]
Lifespan impact	The circadian rhythm influences plant longevity by optimizing energy use and growth cycles, maximizing resource allocation during optimal periods	In animals, circadian misalignment, such as through shift work, is associated with increased aging, obesity, and higher mortality due to disrupted metabolic processes [18]
Molecular clock evolution	The molecular clock in plants has evolved to synchronize key life processes such as photosynthesis, reproduction, and flowering	In animals, the molecular clock evolved to optimize survival through daily cycles of sleep, feeding, and metabolic regulation, crucial for homeostasis [65]

(continued)

Table 2 (continued)

Aspect	Plant circadian rhythm	Animal circadian rhythm
Role in homeostasis	Plants' circadian rhythms help regulate internal balance by ensuring optimal energy use and nutrient allocation throughout the day	In animals, circadian rhythms are central to maintaining homeostasis, particularly in metabolic functions like glucose regulation, fat storage, and energy expenditure [18]
Sensitivity to disruption	In plants, circadian disruption can lead to poor growth, misaligned flowering, and energy inefficiency, affecting survival and reproduction	In animals, disruption of circadian rhythms (e.g., shift work, jet lag) increases the risk of metabolic diseases, such as obesity, insulin resistance, and cardiovascular disease [18]
Impact of artificial light	Artificial light can significantly disrupt plant growth, flowering, and nutrient absorption	Artificial light disrupts sleep, hormone regulation, and metabolism in animals, leading to conditions like insomnia and obesity
Influence of social interactions	Plants do not rely on social cues but can synchronize based on light and temperature	Social interactions in animals can influence circadian rhythms, with group synchronization observed in some species

- Growth and Development—Plant growth patterns, such as hypocotyl elongation and stem elongation, follow circadian regulation. Growth often occurs during specific times of the day to maximize resource utilization.
- Flowering Time Regulation—The timing of flowering is often synchronized with circadian cues to ensure reproductive success. In photoperiodic plants, the circadian clock interacts with light signals to determine the appropriate season for flowering.
- Metabolic Processes—Circadian regulation affects primary and secondary metabolism, influencing nutrient uptake, energy storage, and defense compound production to enhance survival and adaptation.
- Stress Responses—Plants use circadian rhythms to anticipate environmental stresses such as drought, pathogen attacks, and temperature fluctuations, activating protective mechanisms at optimal times.

In summary, circadian rhythms in plants are fundamental for optimizing physiological processes, enhancing survival, and ensuring reproductive success by synchronizing internal functions with external environmental cycles.

References

1. Andronis C, Barak S, Knowles SM, Sugano S, Tobin EM (2008) The clock protein CCA1 and the bZIP transcription factor HY5 physically interact to regulate gene expression in Arabidopsis. Mol Plant 1:58–67
2. Angevin F, Klein EK, Choimet C, Gauffreteau A, Lavigne C, Messéan A, Meynard JM (2008) Modelling impacts of cropping systems and climate on maize cross-pollination in agricultural landscapes: the MAPOD model. Eur J Agron 28:471–484
3. Aschoff J (1989) Temporal orientation: circadian clocks in animals and humans. Anim Behav 37:881–896
4. Atamian HS, Creux NM, Brown RI, Garner AG, Blackman BK, Harmer SL (2016) Circadian regulation of sunflower heliotropism, floral orientation, and pollinator visits. Science 353:587–590
5. Atkins KA, Dodd AN (2014) Circadian regulation of chloroplasts. Curr Opin Plant Biol 21:43–50
6. Baudry A, Kay S (2008) Clock control over plant gene expression. Adv Bot Res 48:69–105
7. Bhadra U, Thakkar N, Das P, Bhadra MP (2017) Evolution of circadian rhythms: from bacteria to human. Sleep Med 35:49–61
8. Boix M, Garcia-Rodriguez A, Castillo L, Miró B, Hamilton F, Tolak S, Pérez A, Monte-Bello C, Caldana C, Henriques R (2024) 40S Ribosomal protein S6 kinase integrates daylength perception and growth regulation in Arabidopsis thaliana. Plant Physiol 195:3039–3052
9. Carré IA (2016) The regulation of circadian rhythms in plants. Encyclopedia of Applied Plant Sciences 2nd ed. 386
10. Cashmore AR (2003) Cryptochromes: enabling plants and animals to determine circadian time. Cell 114:537–543
11. Christie JM, Briggs WR (2001) Blue light sensing in higher plants. J Biol Chem 276:11457–11460
12. Covington MF, Maloof JN, Straume M, Kay SA, Harmer SL (2008) Global transcriptome analysis reveals circadian regulation of key pathways in plant growth and development. Genome Biol 9:1–18
13. Daniel X, Sugano S, Tobin EM (2004) CK2 phosphorylation of CCA1 is necessary for its circadian oscillator function in Arabidopsis. Proc Natl Acad Sci 101:3292–3297
14. David KM, Armbruster U, Tama N, Putterill J (2006) Arabidopsis GIGANTEA protein is post-transcriptionally regulated by light and dark. FEBS Lett 580:1193–1197
15. de Montaigu A, Tóth R, Coupland G (2010) Plant development goes like clockwork. Trends Genet 26:296–306
16. Devlin PF (2002) Signs of the time: Environmental inputs to circadian clocks. J Exp Bot 53:1515–1525
17. Dibner C, Schibler U (2015) Circadian clocks and metabolism. J Intern Med 277:4–10
18. Dibner C, Schibler U (2015) Circadian timing of metabolism in animal models and humans. J Intern Med 277:513–527
19. Doyle MR, Davis SJ, Bastow RM, McWatters HG, Kozma-Bognár L, Nagy F, Millar AJ, Amasino RM (2002) The ELF4 gene controls circadian rhythms and flowering time in Arabidopsis thaliana. Nature 419:74–77
20. Edgar RS, Green EW, Zhao Y, Van Ooijen G, Olmedo M, Qin X, Reddy AB (2012) Peroxiredoxins are conserved markers of circadian rhythms. Nature 485:459–464
21. Edwards KD, Millar AJ (2007) Analysis of circadian leaf movement rhythms in Arabidopsis thaliana. Circadian Rhythm: Methods Protoc, pp 103–113
22. Eriksson ME, Webb AAR (2011) The clock and the stress response. Plant Signal Behav 6:214–216
23. Fekih R, Nefissi R, Miyata K, Ezura H, Mizoguchi T (2009) Roles of circadian clock and histone methylation in the control of floral repressors. Adv Bot Res 50:199–225

24. Fujiwara S, Wang L, Han L, Suh SS, Salomé PA, McClung CR, Somers DE (2008) Post-translational regulation of the Arabidopsis circadian clock through selective proteolysis and phosphorylation of pseudo-response regulator proteins. J Biol Chem 283:23073–23083
25. Green RM, Tingay S, Wang ZY, Tobin EM (2002) Circadian rhythms confer a higher level of fitness to Arabidopsis plants. Plant Physiol 129:576–584
26. Hall A, Gould P (2008) Temperature compensation of the Arabidopsis circadian clock. Comp Biochem Physiol A 150:S153
27. Harmon FG, Kay SA (2003) The F box protein AFR is a positive regulator of phytochrome A-mediated light signaling. Curr Biol 13:2091–2096
28. Heintzen C, Nater M, Apel K, Staiger D (2001) Circadian oscillations of a transcript encoding a germin-like protein that is associated with cell walls in Arabidopsis thaliana. Plant Physiol 125:2078–2088
29. Hsu PY, Harmer SL (2014) Wheels within wheels: the plant circadian system. Trends Plant Sci 19:240–249
30. Hubbard KE, Webb AAR (2011) Circadian rhythms in stomata: physiological and molecular aspects. Plant Physiol 155:17–27
31. Iijima M, Matsushita N (2011) A circadian and an ultradian rhythm are both evident in root growth of rice. J Plant Physiol 168:2072–2080
32. Imaizumi T (2010) Arabidopsis circadian clock and photoperiodism: time to think about location. Curr Opin Plant Biol 13:83–89
33. Kevei E, Gyula P, Fehér B, Tóth R, Viczián A, Kircher S, Rea D, Dorjgotov D, Schäfer E, Millar AJ, Kozma-Bognár L (2007) Arabidopsis thaliana circadian clock is regulated by the small GTPase LIP1. Curr Biol 17:1456–1464
34. Kim JR, Bae WS, Yoon Y, Cho KH (2007) Topological differences in plant and animal circadian regulatory networks. Biophys J 92:L01–L03
35. Kim JA, Yang TJ, Kim JS, Park JY, Kwon SJ, Lim MH, Jin M, Lee SC, Lee SI, Choi BS, Um SH (2007) Isolation of circadian-associated genes in Brassica rapa by comparative genomics with Arabidopsis thaliana. Mol Cells 23:145–153
36. Kiyota M, Numayama N, Goto K (2006) Circadian rhythms of the L-ascorbic acid level in Euglena and spinach. J Photochem Photobiol, B 84:197–203
37. Lau OS, Huang X, Charron JB, Lee JH, Li G, Deng XW (2011) Interaction of Arabidopsis DET1 with CCA1 and LHY in mediating transcriptional repression in the plant circadian clock. Mol Cell 43:703–712
38. Marceau A, Loubet B, Andrieu B, Durand B, Foueillassar X, Huber L (2011) Modelling diurnal and seasonal patterns of maize pollen emission in relation to meteorological factors. Agric For Meteorol 151:11–21
39. Mas P (2008) Circadian clock function in Arabidopsis thaliana: time beyond transcription. Trends Cell Biol 18:273–281
40. Mas P, Yanovsky MJ (2009) Time for circadian rhythms: plants get synchronized. Curr Opin Plant Biol 12:574–579
41. Matsushika A, Makino S, Kojima M, Yamashino T, Mizuno T (2002) The APRR1/TOC1 quintet implicated in circadian rhythms of Arabidopsis thaliana: II. Characterization with CCA1-overexpressing plants. Plant Cell Physiol 43:118–122
42. McClung CR (2001) Circadian rhythms in plants. Annu Rev Plant Biol 52:139–162
43. McClung CR (2011) The genetics of plant clocks. New Phytol 190:600–613
44. McClung CR, Gutierrez RA (2010) Network news: prime time for systems biology of the plant circadian clock. Curr Opin Genet Dev 20:588–598
45. McClung CR, Salomé PA, Michael TP (2002) The Arabidopsis circadian system. Arab Book/ Am Soc Plant Biol 1:1–18
46. McWatters HG, Devlin PF (2011) Timing in plants–a rhythmic arrangement. FEBS Lett 585:1474–1484
47. Mizuno T (2004) Plant response regulators implicated in signal transduction pathways essential for circadian rhythms. Curr Opin Plant Biol 7:595–598

48. Moran N (2015) Rhythmic leaf movements: physiological and molecular aspects. Rhythm Plants: Dyn Responses Dyn Environ, pp 57–95
49. Nagel DH, Kay SA (2012) Complexity in the wiring and regulation of plant circadian networks. Curr Biol 22:R648–R657
50. Nakamichi N, Kusano M, Fukushima A, Kita M, Ito S, Yamashino T, Saito K, Sakakibara H, Mizuno T (2009) Transcript profiling of an Arabidopsis PSEUDO RESPONSE REGULATOR arrhythmic triple mutant reveals a role for the circadian clock in cold stress response. Plant Cell Physiol 50:447–462
51. Pokhilko A, Mas P, Millar AJ (2013) Modelling the widespread effects of TOC1 signalling on the plant circadian clock and its outputs. BMC Syst Biol 7:1–12
52. Prombona A, Argyroudi-Akoyunoglou J (2004) Diverse signals synchronise the circadian clock controlling the oscillations in chlorophyll content of etiolated Phaseolus vulgaris leaves. Plant Sci 167:117–127
53. Roden LC, Carré IA (2001) The molecular genetics of circadian rhythms in Arabidopsis. Semin Cell Dev Biol 12(4):305–315
54. Seo PJ, Mas P (2015) STRESSing the role of the plant circadian clock. Trends Plant Sci 20:230–237
55. Shin BS, Lee JH, Lee JH, Jeong HJ, Yun CH, Kim JK (2004) Circadian regulation of rice (Oryza sativa L.) CONSTANS-like gene transcripts. Mol Cells 17:10–16
56. Solomon OL, Berger DK, Myburg AA (2010) Diurnal and circadian patterns of gene expression in the developing xylem of Eucalyptus trees. S Afr J Bot 76:425–439
57. Southern MM, Millar AJ (2005) Circadian genetics in the model higher plant, Arabidopsis thaliana. Methods Enzymol 393:23–35
58. Staiger D, Green R (2011) RNA-based regulation in the plant circadian clock. Trends Plant Sci 16:517–523
59. Stratmann T, Más P (2008) Chromatin, photoperiod and the Arabidopsis circadian clock: a question of time. Semin Cell Dev Biol 19(6):554–559
60. Sullivan JA, Deng XW (2003) From seed to seed: the role of photoreceptors in Arabidopsis development. Dev Biol 260:289–297
61. Thines B, Harmon FG (2011) Four easy pieces: mechanisms underlying circadian regulation of growth and development. Curr Opin Plant Biol 14:31–37
62. Travlos IS, Liakopoulos G, Karabourniotis G, Fasseas C, Karamanos AJ (2008) Circadian leaflet movements of Tylosema esculentum (Burch) A. Schreib, and the abolishment of these diurnal movements by potassium deficiency. J Arid Environ 72:1745–1750
63. Troncoso-Ponce MA, Mas P (2012) Newly described components and regulatory mechanisms of circadian clock function in Arabidopsis thaliana. Mol Plant 5:545–553
64. Van Hout R, Chamecki M, Brush G, Katz J, Parlange MB (2008) The influence of local meteorological conditions on the circadian rhythm of corn (Zea mays L.) pollen emission. Agric For Meteorol 148:1078–1092
65. Vitaterna MH, Takahashi JS, Turek FW (2001) Overview of circadian rhythms. Alcohol Res Health 25:85
66. Wang H, Wang H (2015) Phytochrome signaling: time to tighten up the loose ends. Mol Plant 8:540–551
67. Yakir E, Hassidim M, Melamed-Book N, Hilman D, Kron I, Green RM (2011) Cell autonomous and cell-type specific circadian rhythms in Arabidopsis. Plant J 68:520–531
68. Yang M, Lin W, Xu Y, Xie B, Yu B, Chen L, Huang W (2024) Flowering-time regulation by the circadian clock: from Arabidopsis to crops. The Crop Journal 12:17–27
69. Yanovsky MJ, Kay SA (2001) Signaling networks in the plant circadian system. Curr Opin Plant Biol 4:429–435
70. Yari Kamrani Y, Shomali A, Aliniaeifard S, Lastochkina O, Moosavi-Nezhad M, Hajinajaf N, Talar U (2022) Regulatory role of circadian clocks on ABA production and signaling, stomatal responses, and water-use efficiency under water-deficit conditions. Cells 11:1154

71. Yu JW, Rubio V, Lee NY, Bai S, Lee SY, Kim SS, Liu L, Zhang Y, Irigoyen ML, Sullivan JA, Zhang Y (2008) COP1 and ELF3 control circadian function and photoperiodic flowering by regulating GI stability. Mol Cell 32:617–630
72. Zhao H, Xu D, Tian T, Kong F, Lin K, Gan S, Zhang H, Li G (2021) Molecular and functional dissection of EARLY-FLOWERING 3 (ELF3) and ELF4 in Arabidopsis. Plant Sci 303:110786

Chapter 2
Plant Response to Environmental Periodicity: Diurnal Rhythms

Prasann Kumar, Shipa Rani Dey, Joginder Singh, and Neha Gogia

1 Introduction

Environmental periodicity is the periodicity of ecological factors, such as light, temperature, and humidity, according to specific time intervals. Those cyclic changes, dictated by the Earth's spinning and its path around the sun, produce rhythms, the daily pattern of day and night, seasons, and variations in tides. In biological systems, organisms have evolved biological rhythms and internal processes that help predict and adapt to these recurring variations [3]. Circadian rhythms' endogenous 24-h cycles are essential in entraining various physiological and behavioral processes. Active venturing into the surrounding atmosphere is fundamental to survival and reproductive success. Organisms adjust their metabolic, developmental, and behavioral methods to environmental signals to maximize energy allocation, resource use, and ecological interactions. Plants have been known, for example, to fine-tune photosynthesis, flowering, and stomatal opening according to daily light cycles to optimize efficiency while limiting water loss.

Outside of biological contexts, environment-wide periodicity has profound implications for agriculture, conservation, and climate science. Timing periodic changes allows seasonal crop yield predictions, pest population tracking, and biodiversity

P. Kumar (✉)
Department of Agronomy, School of Agriculture, Lovely Professional University, Phagwara, Punjab, India
e-mail: prasann0659@gmail.com

S. R. Dey
Department of Agronomy, Faculty of Agricultural Sciences, GLA University, Uttar Pradesh, Mathura, Chaumuhan, India

J. Singh
Department of Botany, Nagaland University, Nagaland, India

N. Gogia
New Haven, CT, USA

A. Kothari Chhajer and V. Singh Gour (eds.), *Introduction to Rhythms in Plants*, Studies in Rhythm Engineering, https://doi.org/10.1007/978-981-95-3592-7_2

management. Human activity that disrupts natural periodicity, such as light pollution and climate change, can have cascading effects on ecosystems. Thus, environmental periodicity is crucial for the measures to be consistent with the balance of human development and ecosystem stability [8].

Daily rhythms of biological activity, known as circadian rhythms, are basic 24-h cycles regulated by molecular clocks that can be found in many forms of life, even those that do not have eyes, such as plants. Zeitgebers (or time-giving cues) entrained these rhythms in the external environment, like light and temperature. Daily cycles of light and dark control a myriad of physiological and developmental processes in plants, ensuring that these processes occur at times aligned with the most beneficial environmental conditions.

Photosynthesis is one of the most critical processes regulated by daily rhythms. During the day, plants switch on their photosynthetic machinery and absorb light, which they store as chemical energy. At the same time, stomata are open to exchange, which allows absorption of CO_2 while minimizing the water loss [4]. The circadian clock regulates the process. It is governed by the cycling of energy balance in the plant, with downregulation of metabolism at night when resources are being conserved and increased preparation for the next day's challenges.

Daily rhythms regulate other processes, too, including growth and flowering. For instance, photoperiodism, which governs how plants determine the response of the amount of sunlight to that of darkness, is critical for regulating the timing of flowering. Short-day plants like chrysanthemums flower when days are shorter, while long-day plants like spinach bloom in prolonged sunlight. Such coordination of flowering allows flowering to synchronize with the most favorable state for reproduction, animal pollination, and its consequences.

Diurnal rhythms have been shown to influence stress responses as well. Plant circadian-regulated systems also allow plants to mitigate the effects of abiotic stresses, such as drought and heat, by predicting and easing their impact. For instance, genes involved in stress tolerance often peak at certain times of day, preparing the plant for anticipated environmental shifts. [11]. These adaptation mechanisms underscore the significance of circadian rhythms, which are vital to plant health, productivity, and resilience.

The plants' responses to light and dark cycles have long been subject to academic study, with their origins stretching back thousands of years. However, when other flowers opened and closed day by day, early naturalists observed their quotidian opening and closing, including the "sleep movements" of the *Mimosa pudica*. But it was not until the eighteenth century that organized scientific inquests commenced, led by the enterprising efforts of Jean-Jacques d'Ortous de Mairan. In 1729, de Mairan found that the leaf movements of *Mimosa* plants continued under constant darkness, thus showing that an internal biological clock governed this behavior.

Later studies of plant rhythms by Carl Linnaeus and Charles Darwin built on these discoveries. One of Linnaeus's ideas was for a "flower clock," a garden design where flowers bloom and close at specific times of day, by their circadian rhythms. Darwin's experiments, recounted in The Power of Movement in Plants (1880), addressed phototropism and heliotropism, particularly how plants orient themselves toward the light and respond to its intensity.

In the twentieth century, much of the molecular basis of circadian rhythms was understood. The discovery of photoreceptors such as phytochromes and cryptochromes made understanding how plants sense light possible. The genetic basis underlying circadian regulation was provided by identifying the first clock genes, such as TOC1 (TIMING OF CAB EXPRESSION 1) and CCA1 (CIRCADIAN CLOCK ASSOCIATED 1). These developments paved the way for modern chronobiology, enabling researchers to detail environmental cues and internal clocks [5].

Understanding daily rhythms is becoming widely recognized as a critical consideration for those attempting to tackle challenges in agroecological sustainability. Agricultural control is a significant objective to maximize crop output and resource effectiveness. Plant circadian rhythms control growth rate, nutrient uptake, and stress tolerance. For example, altering flowering time, i.e., photoperiod sensitivity, could extend growing seasons or help crops adapt to varying environments [20].

Daily rhythms are also crucial in pest and pathogen management. Many crop pests exhibit circadian rhythms in feeding and mating so that interventions can be concentrated during peak activity. Knowing when such responses occur can enhance pest control strategies and, thus, reduce chemical application.

Many ecological interactions are regulated by daily rhythms, e.g., pollination and herbivory. Disruptions to these rhythms, courtesy of artificial light at night or climate-induced phenological shifts, can destabilize ecosystems. The implications of understanding these dynamics are relevant to conservation plans that mitigate human-derived effects.

And as climate change forces our surroundings to change faster than our plants can evolve, understanding how plants align with their environments becomes critical. Changes in temperature and precipitation can trigger a desynchronization of plant rhythms and their historical environment that can have consequences for productivity and biodiversity. Including chronobiology in climate models could yield more accurate predictions of how ecosystems will react, allowing for better-formed adaptations [2, 6].

Overall, these concepts of periodicity across environmental conditions and daily rhythms build the foundations for how we approach plant biology and ecological systems. Examining these rhythms, from historical approaches to contemporary uses across various practices, including agriculture and sustainability, reveals the interconnectedness of life and their ambient and cyclical presence in the environment. With threat levels rising worldwide, who will apply our understanding of these natural cycles to foster resilience and a balance between human behavior and nature?

2 The Biological Clock: Circadian Rhythms in Plants

Circadian rhythms are endogenously generated, free-running, and persistent oscillations with a period of approximately 24 h, and they are found in a broad range of biological entities, including the plant kingdom. These circadian rhythms enable organisms to anticipate and prepare for circadian variations internal to their surroundings, thus granting a significant selective advantage. Circadian rhythms regulate various processes beyond diurnal fluctuations in gene expression, including growth, flowering, and photosynthesis, among many physiological processes in plants. The circadian system has three interconnected components: the central oscillator, which acts as the core mechanistic generator of rhythmicity; input pathways allowing environmental light cues to synchronize the oscillator; and output signals allowing oscillatory mechanistic drives physiological and molecular responses [12, 21].

The plant master oscillator comprises a complex network of interlinked transcriptional and translational feedback loops. A network of clock-associated genes and their protein products that feed back to regulate their expression forms the basis of this system. These consist of, for instance, CIRCADIAN CLOCK ASSOCIATED 1 (CCA1), LATE ELONGATED HYPOCOTYL (LHY), and TIMING OF CAB EXPRESSION 1 (TOC1). At the heart of generating the circadian rhythm of these components is a core loop comprised of the morning-expressed transcription factors CCA1 and LHY that repress the expression of TOC1 during the day. TOC1 is activated in the evening and suppresses the expression of CCA1 and LHY at night. This feedback loop leads to a robust circadian oscillation. Some other forward genes, including GIGANTEA (GI), EARLY FLOWERING 3(ELF3), and PSEUDO-RESPONSE REGULATORS (PRRs), were also linked from the core loop, respectively, making the oscillator complex, stable, and robust [18, 19].

Input pathways are critical to entraining or synchronizing the central oscillator with external environmental cues. In plants, light and temperature are external cues. Light input is perceived by the photoreceptors phytochromes (Phy) and cryptochromes (Cry), and the F-box proteins ZEITLUPE (ZTL) and its relative, photoreceptor-interacting (ZTL), act in light-regulated degradation of clock proteins, including TOC1, PRR5. Light modifies the expression and activity of clock genes to synchronize internal rhythms with the external light/dark cycle. Temperatures also influence circadian clocks through thermosensitive transcriptional regulation and protein stability pathways, and it has been suggested that circadian clocks can maintain rhythm fidelity under disparate thermal environments [19].

The circadian system generates signals that regulate downstream physiological processes. These cues are communicated through clock-controlled genes (CCGs), the expression of which is rhythmically driven by the central oscillator. These CCGs are implicated in diverse processes, such as photosynthesis, hormone signaling, and stress responses, to adjust plant performance to the diel cycles. For example, the circadian regulation of the CAB (CHLOROPHYLL A/B-BINDING) gene expression is synchronized with photonic activity to optimize energy harvest. Similarly, circadian regulation of stomatal opening enhances water-use efficiency by controlling transpiration according to the light environment [1, 22].

The core clock genes, such as CCA1, LHY, and TOC1, underlie the transcriptional feedback loops. CCA1 and LHY are MYB transcription factors that bind to evening elements (EE) in TOC1 and other evening-expressed gene promoters and repress their expression. TOC1 then inhibits the expression of the morning-expressed genes, thereby closing the loop. PRRs such as PRR7 and PRR9 are transcriptional repressors of CCA1 and LHY, adding a second layer of regulation. Even when the environment stays the same, the interaction of key factors causes ongoing rhythms or cycles that don't weaken over time. This shows how the internal clock can operate freely while staying in sync with an organized and balanced system (Table 1).

The expression of morning genes is also controlled by the evening complex (EC), which includes PRR9, ERF5, and LUX. This complex helps the plant smoothly shift from evening to morning phases. They also include post-transcriptional and post-translational modifications necessary for clock regulation [17, 24]. Differential splicing between clock gene transcripts, phosphorylation of clock proteins, and selective degradation of proteins through a ubiquitin-mediated process further regulate clock activity and fine-tune the circadian rhythm.

The interaction pathways of circadian rhythms and daily cycles are dynamic and mediated by phase shifting and entrainment processes. In contrast, phase shifting is a forward or delayed circadian oscillation phase induced by environmental signals. For example, light exposure during the subjective night (e.g., the dark phase of the circadian cycle) can result in phase advances or phase delays depending on the timing of the light signal. Photoreceptors that mediate light-induced stabilization (Phy) or degradation (Cry) of clock proteins regulate this light sensitivity. Cycles of temperature also cause phase shifts: warm temperatures move the clock forward while cool temperatures take it back. These modifications ensure that the circadian clock stays in sync with the environment.

Entrainment is the process by which the circadian clock is reset in alignment with periodic environmental signals, such as the light/dark cycle or daily fluctuations in temperature. These signal inputs are photoreceptors that detect the quality, quantity, and duration of light in plants to synchronize the clock. For example, phytochrome initiated by red light can activate downstream signaling, stabilizing the PRR protein and resetting the clock phase. Five central photoreceptors mediate the responsiveness of the circadian clock to light: cryptochromes (Cry1 and Cry2) and zenith-like proteins (Zeal and ZTL) are activated by blue light to induce the degradation of clock proteins and to modulate the pacemaker. On the other hand, temperature entrainment refers to the transcriptional and translational activity changes of the clock genes in response to thermal changes. The coordination of light and temperature signals enables tight circadian rhythms in the surrounding milieu of the plants. For example, the circadian regulation of photosynthesis allows for the maximal carbon fixation to occur during the day when light energy is available circadian gating of epidermal stomatal opening balances water loss with carbon fixation demand for efficient water usage and improved photosynthesis. Furthermore, hormone signaling pathways such

Table 1 Biological clock: circadian rhythms in plants

Aspect	Key features	Molecular basis	Environmental interactions	Agricultural relevance	Examples	Future research directions
Photosynthesis	Peak photosynthetic activity aligned with daylight hours	Genes: TOC1, CCA1, LHY; Proteins: Rubisco	Light intensity, photoperiod	Enhanced biomass production and resource use efficiency	Arabidopsis thaliana, rice	Genetic manipulation to optimize photosynthetic timing for increased yield
Stomatal regulation	Stomata open during the day and close at night to optimize gas exchange	Genes: FLC, ZEITLUPE; Hormones: ABA	Light cycles, humidity	Improved water use efficiency and drought tolerance	Maize, soybean	Developing crops with adjustable stomatal rhythms for arid environments
Hormone production	Timing of auxin, cytokinin, and ABA peaks for growth and stress responses	Genes: AUX/IAA family, ABAR (ABA receptor)	Temperature, light–dark cycles	Better growth and stress management	Wheat, tomato	Fine-tuning hormone cycles to improve stress resilience and growth rates
Stress responses	Circadian control of defense gene expression during specific times	Pathways: JA, SA, ROS	Temperature shifts, photoperiod	Enhanced resistance to pests and diseases	Tobacco, barley	Identifying optimal timing for pesticide application based on defense rhythm
Nutrient uptake	Diurnal timing of nitrogen assimilation and nutrient transport	Pathways: NRT1.1, NR	Light–dark phases, soil temperature	Increased fertilizer efficiency and reduced environmental impact	Arabidopsis, rice	Synchronizing nutrient application with uptake cycles to maximize efficiency
Flowering timing	Regulation of flowering by photoperiod and circadian clock	Genes: CO, FT, SOC1	Day length, temperature	Optimal flowering time for higher pollination success and yield	Barley, Arabidopsis	Engineering flowering times for seasonal adaptation

(continued)

Table 1 (continued)

Aspect	Key features	Molecular basis	Environmental interactions	Agricultural relevance	Examples	Future research directions
Growth and cell division	Cell elongation and division peaks during specific times of the day	Pathways: TOR signaling, Expansin proteins	Light cycles, temperature	Accelerated growth rates and improved resource allocation	Corn, lettuce	Integrating circadian cues into Growth models for controlled environment farming
Root development	Circadian modulation of root growth and branching for efficient nutrient uptake	Genes: ARF family, HY5	Soil temperature, moisture levels	Improved nutrient and water uptake	Soybean, Arabidopsis	Studying root clock interactions for optimizing underground resource acquisition
Metabolite production	Timing of secondary metabolite synthesis for defense and pollinator attraction	Pathways: MYB transcription factors	Light cycles, biotic stress	Enhanced production of beneficial compounds for plant and human use	Mint, basil	Unlocking timing mechanisms to maximize secondary metabolite yields
Circadian recalibration	Adjustments of circadian clocks under shifting environmental conditions	Genes: PRR family, ELF3	Changing photoperiods, extreme weather	Climate-resilient crops with adaptable circadian clocks	Sunflower, rice	Designing crops that can recalibrate rhythms for climate adaptation
Seed germination	Regulation of germination timing to match optimal environmental conditions	Hormones: GA, ABA	Light, temperature fluctuations	Improved germination rates and seedling establishment	Barley, Arabidopsis	Manipulating germination clocks for enhanced crop establishment in suboptimal conditions

(continued)

Table 1 (continued)

Aspect	Key features	Molecular basis	Environmental interactions	Agricultural relevance	Examples	Future research directions
Carbon assimilation	Diurnal regulation of carbon storage and mobilization for energy	Genes: TPS1, SnRK1	Light–dark transitions	Increased carbon use efficiency and growth	Sugarcane, maize	Exploring clock-based carbon partitioning for bioenergy crops
Photomorphogenesis	Light-driven developmental changes synchronized with circadian rhythms	Genes: PHY, PIF3, COP1	Light quality, intensity	Enhanced plant architecture for optimal light capture	Arabidopsis, mustard	Developing crops with tailored photomorphogenic responses for dense planting systems
Leaf movements	Nyctinastic leaf movements are coordinated with light availability	Pathways: Pulvinar motor cells, K + channels	Light–dark cycles	Maximized photosynthetic efficiency and reduced herbivory risk	Mimosa, beans	Investigating the molecular basis of leaf movements to improve plant adaptability
Fruit ripening	Timing of ripening to synchronize with environmental and developmental cues	Genes: RIN, NOR, CNR	Temperature, ethylene signaling	Improved harvest timing and fruit quality	Tomato, banana	Controlling ripening schedules for extended shelf life and market adaptability

as those involved in abscisic acid (ABA) and Jasmonic acid (JA) signaling have been shown to contain circadian clock modifiers, allowing for the coordination of growth and stress responses with environmental cues [16, 23].

Plant circadian rhythms are sophisticated communication systems that involve central oscillators, input pathways, and output signals, enabling the temporal regulation of various physiological processes in response to daily environmental variations. Elements of the core clock, such as CCA1, LHY, and TOC1, underpin the circadian oscillator, whereas input pathways, which consist of photoreceptors and temperature sensors, entrain the clock to external cues. These output signals translate rhythmic information into temporal regulation of downstream processes, ultimately enhancing plant adaptation and fitness. These systems exert a crucial evolutionary role in plant biology, with the co-action of circadian and diurnal rhythms [13, 14].

3 Photoreceptors and Light Signaling Pathways

Once light strikes the leaf, photoreceptors are specialized molecules that perceive the light, including its intensity, direction, wavelength, and duration. These photoreceptors are necessary sensors, enabling plants to adjust to their surroundings. Prominent photoreceptors, including the phytochromes, cryptochromes, phototropins, and UVR8, mediate diverse yet integrated roles in light signal capture and its transduction.

Phytochromes are a group of red and far-red light photoreceptors in two interconvertible forms: Pr, which absorbs red light, and Pfr, which absorbs far-red light. When Pr is exposed to red light, it converts to Pfr, the active form, activating subsequent signaling cascades. This reversible transformation enables phytochromes to act as molecular photo-switches controlling seed germination, shade avoidance, and flowering. For example, phytochromes mediate photomorphogenesis by modulating gene expression based on the ratio of red and far-red lights perceived, which also helps plants adapt to light environments [10, 15].

On the other hand, cryptochromes are photoreceptors sensitive to blue light and play an essential part in circadian rhythm regulation. Plant and animal cryptochromes bind blue and ultraviolet-A (UV-A) through their flavin adenine dinucleotide (FAD) chromophores. Cryptochromes are essential in plants' photoperiodic flowering, stem elongation, and stomatal opening. Cryptochromes enable plants to entrain their physiological responses to the day-night cycle by modulating the transcription of specific genes.

Another class of blue light photoreceptors is the phototropins, primarily responsible for phototropism, the directional growth of plants toward or away from a light source. These proteins have two light, oxygen, or voltage (LOV) domains, which sense blue light, leading to kinase activity associated with the proteins. This causes the redistribution of auxin, a plant hormone, and promotes asymmetric cell elongation. In addition to directing chloroplast movement for maximum light capture and stomatal opening, phytotropins synchronize gas exchange to ensure optimal conditions for photosynthesis [7, 9, 25].

UVR8 is the photoreceptor of UV-B (ultraviolet-B) light and is the only known mechanism by which plants protect themselves from potentially damaging UV-B radiation. Indeed, while many known photoreceptors function via chromophores, UVR8 directly absorbs UV-B light via its tryptophan residues. The monomeric form of UVR8 then interacts with the signaling protein COP1 in response to UV-B stimuli, leading to the expression of various genes involved in UV defense, including genes encoding flavonoids and other photoprotective compounds [6, 19, 23]. Balancing growth promotion and stress alleviation creates a catch-22 for plant health, such that UV-B acts as a double-edged sword, and UVR8 helps set the tone for what kind of plant behavior comes next (Table 2).

Plants respond to environmental factors, and the perception and interpretation of light quality are critical for plant growth and development. Different photoreceptors mediate unique physiological responses depending on the light wavelength (red, far-red, blue, and UV). Perception of light at specific wavelengths is essential for seedling establishment, shade avoidance, flowering, and adaptation to stress.

The shade avoidance response is controlled by red and far-red light, which phytochromes detect. This is because red light is absorbed by chlorophyll, and far-red light is reflected in dense plant canopies, which decreases the ratio of red to far-red light. In response, phytochromes trigger stem and petiole elongation in plants so that they grow over other plants competing for light. This mechanism underscores how light quality cues determine plant architecture and resource allocation [14, 15, 24].

Cryptochrome and phototropin implants, sensitive to blue light, control various processes, including phototropism, chloroplast movement, and stomatal behavior. Cryptochromes mediate blue light-induced hypocotyl elongation inhibition, guaranteeing seedling photomorphogenesis. At the same time, phototropins adjust chloroplast positioning to enhance photosynthesis in low-light conditions and minimize light damage in high-light conditions.

In plant biology, UV light, particularly UV-B, has a dual role. Although UV-B at high intensity can damage multiple cell structures, low levels are sensed by the UVR8 photoreceptor to induce stress-resilience pathways. This mechanism involves the synthesis of up-regulated UV-absorbing pigments, DNA repair enzymes, and antioxidants, thereby increasing the plant's tolerance to environmental stressors [9, 12, 16].

The circadian clock is an internal timing system that lets the plant predict daily environmental changes. Light, a principal zeitgeber (time cue), entrains the circadian clock by resetting it to the exogenous light–dark cycle. Photoreceptors, most notably the phytochromes and cryptochromes, are crucial to this synchronization as they transduce light signals into molecular oscillations.

Aside from clock regulation, light affects the expression of genes related to photosynthesis. Photoreceptor signals relayed by blue and red-light signaling pathways trigger transcription factors that promote the expression of photosynthetic proteins, including ribulose-1,5-bisphosphate carboxylase/oxygenase (Rubisco) and light-harvesting chlorophyll-binding proteins. These adjustments allow plants to leverage energy from light to fix carbon in a manner that matches metabolic function to environmental context.

Table 2 Different types of photoreceptors and their role

Photoreceptor type	Light spectrum sensitivity	Molecular structure	Signaling pathway components	Biological functions	Environmental interaction	Agricultural implications	Examples	Future research directions
Phytochromes	Red (660 nm), Far-red (730 nm)	Chromophore: Phytochromobilin; Domains: PAS, GAF, PHY	PHYB, PHYA, PIFs, COP1, HY5	Photomorphogenesis, shade avoidance, and seed germination	Light quality (red:far-red ratio), photoperiod	Engineering shade-tolerant crops; enhancing light-use efficiency	Arabidopsis, rice	Studying mechanisms to optimize red: far-red signaling for increased biomass and yield
Cryptochromes	Blue (400–500 nm)	Chromophore: Flavin adenine dinucleotide (FAD); PHR domain	CRY1, CRY2, SPA1, COP1, HY5	Flowering, circadian rhythm regulation	Light intensity, photoperiod	Improving flowering synchronization; enhancing circadian stability	Arabidopsis, tomato	Development of cryptochrome variants for controlled flowering and photoperiod adaptation
Phototropins	Blue (400–500 nm)	Chromophore: Flavin mononucleotide (FMN); LOV domains	NPH1, NPL1, PKS1, ABCB19	Phototropism, stomatal opening, and chloroplast movement	Light direction, intensity	Enhancing light capture efficiency through optimized leaf positioning	Arabidopsis, maize	Exploring phototropin pathways to improve light-harvesting in dense planting systems
UVR8	UV-B (280–315 nm)	Chromophore: Tryptophan residues; No external chromophore	UVR8, COP1, HY5	UV protection, flavonoid biosynthesis, stress response	UV-B radiation, high light intensity	Engineering crops with improved UV tolerance and secondary metabolite production	Arabidopsis, cucumber	Investigating UVR8's role in stress mitigation and secondary metabolite enhancement

(continued)

Table 2 (continued)

Photoreceptor type	Light spectrum sensitivity	Molecular structure	Signaling pathway components	Biological functions	Environmental interaction	Agricultural implications	Examples	Future research directions
ZEITLUPE (ZTL)	Blue (400–500 nm)	LOV domain, F-box protein	ZTL, FKF1, LKP2, GI, TOC1	Circadian clock regulation	Light intensity, temperature	Improving circadian alignment for growth and productivity	Arabidopsis, rice	Exploring ZTL pathways for optimizing growth in varying environmental conditions
LOV-kinases	Blue (400–500 nm)	LOV domain linked to kinase activity	Phototropins, FKF1	Light signaling in diverse developmental processes	Light intensity, daylength	Enhancing adaptability of plants to low-light environments	Arabidopsis, pea	Creating synthetic LOV-kinase systems for precise control of light responses
Flavoproteins	Blue, Green, UV	Chromophores: Flavins and pterins	Phototropins, cryptochromes	Multiple roles in light perception and signaling	Light intensity, quality	Broad applications in improving stress tolerance and photosynthesis efficiency	Spinach, Arabidopsis	Expanding knowledge of noncanonical flavoprotein functions in plant signaling
Phytochromes-Cryptochromes Interaction	Red, Far-red, Blue	Combined chromophores of phytochromes and cryptochromes	PHYB-CRY1 complex, HY5	Integrative light signaling for optimized growth	Light quality, photoperiod	Synergistic manipulation for better plant architecture and flowering regulation	Arabidopsis, sunflower	Understanding cross-talk mechanisms to integrate multiple light signals for superior crop performance

(continued)

Table 2 (continued)

Photoreceptor type	Light spectrum sensitivity	Molecular structure	Signaling pathway components	Biological functions	Environmental interaction	Agricultural implications	Examples	Future research directions
Opsins	Green, Blue, UV	Retinal-binding proteins	Opsin homologs in algae	Photosensory signaling in primitive plants	Light quality, aquatic conditions	Potential bioengineering applications for land plants	Chlamydomonas reinhardtii	Adapting opsin-based signaling to terrestrial crops for novel light-driven processes
Redox-based sensors	Multiple spectrums	Reactive oxygen species (ROS) as signaling molecules	ROS sensors, HY5, MAPKs	Stress signaling, light acclimation	High light intensity, oxidative stress	Developing ROS-sensitive pathways for enhanced stress tolerance	Arabidopsis, wheat	Targeting redox signaling to improve light stress management in fluctuating environments
Cyanobacterial phytochromes	Red, far-red	Linear tetrapyrrole chromophores	Cph1, PixJ, RcaE	Photosensory adaptation in cyanobacteria	Light quality, aquatic conditions	Utilizing cyanobacterial photoreceptors for synthetic biology in crop improvement	Synechocystis sp.	Translating cyanobacterial light-sensing systems for advanced crop bioengineering
Photoreceptor networks	All spectrums	Integrated pathways of multiple photoreceptors	PHY-CRY-UVR8 networks, HY5, COP1, SPA	Coordinated developmental and adaptive responses	Light intensity, quality, and spectrum	Comprehensive control of light-driven processes to boost crop resilience and productivity	Arabidopsis, barley	Decoding complex photoreceptor networks for precise agricultural applications
Photoreceptor Type	Light spectrum sensitivity	Molecular structure	Signaling pathway components	Biological functions	Environmental interaction	Agricultural implications	Examples	Future research directions

(continued)

Table 2 (continued)

Photoreceptor type	Light spectrum sensitivity	Molecular structure	Signaling pathway components	Biological functions	Environmental interaction	Agricultural implications	Examples	Future research directions
Phytochromes	Red (660 nm), far-red (730 nm)	Chromophore: phytochromobilin; domains: PAS, GAF, PHY	PHYB, PHYA, PIFs, COP1, HY5	Photomorphogenesis, shade avoidance, seed germination	Light quality (red:far-red ratio), photoperiod	Engineering shade-tolerant crops; enhancing light-use efficiency	Arabidopsis, rice	Studying mechanisms to optimize red:far-red signaling for increased biomass and yield
Cryptochromes	Blue (400–500 nm)	Chromophore: flavin adenine dinucleotide (FAD); PHR domain	CRY1, CRY2, SPA1, COP1, HY5	Flowering, circadian rhythm regulation	Light intensity, photoperiod	Improving flowering synchronization; enhancing circadian stability	Arabidopsis, tomato	Development of cryptochrome variants for controlled flowering and photoperiod adaptation
Phototropins	Blue (400–500 nm)	Chromophore: Flavin mononucleotide (FMN); LOV domains	NPH1, NPL1, PKS1, ABCB19	Phototropism, stomatal opening, chloroplast movement	Light direction, intensity	Enhancing light capture efficiency through optimized leaf positioning	Arabidopsis, maize	Exploring phototropin pathways to improve light-harvesting in dense planting systems
UVR8	UV-B (280–315 nm)	Chromophore: Tryptophan residues; No external chromophore	UVR8, COP1, HY5	UV protection, flavonoid biosynthesis, stress response	UV-B radiation, high light intensity	Engineering crops with improved UV tolerance and secondary metabolite production	Arabidopsis, cucumber	Investigating UVR8's role in stress mitigation and secondary metabolite enhancement

(continued)

Table 2 (continued)

Photoreceptor type	Light spectrum sensitivity	Molecular structure	Signaling pathway components	Biological functions	Environmental interaction	Agricultural implications	Examples	Future research directions
ZEITLUPE (ZTL)	Blue (400–500 nm)	LOV domain, F-box protein	ZTL, FKF1, LKP2, GI, TOC1	Circadian clock regulation	Light intensity, temperature	Improving circadian alignment for growth and productivity	Arabidopsis, rice	Exploring ZTL pathways for optimizing growth in varying environmental conditions
LOV-Kinases	Blue (400–500 nm)	LOV domain linked to kinase activity	Phototropins, FKF1	Light signaling in diverse developmental processes	Light intensity, daylength	Enhancing adaptability of plants to low-light environments	Arabidopsis, pea	Creating synthetic LOV-kinase systems for precise control of light responses
Flavoproteins	Blue, Green, UV	Chromophores: Flavins and pterins	Phototropins, cryptochromes	Multiple roles in light perception and signaling	Light intensity, quality	Broad applications in improving stress tolerance and photosynthesis efficiency	Spinach, Arabidopsis	Expanding knowledge of noncanonical flavoprotein functions in plant signaling
Phytochromes-Cryptochromes Interaction	Red, Far-red, Blue	Combined chromophores of phytochromes and cryptochromes	PHYB-CRY1 complex, HY5	Integrative light signaling for optimized growth	Light quality, photoperiod	Synergistic manipulation for better plant architecture and flowering regulation	Arabidopsis, sunflower	Understanding cross-talk mechanisms to integrate multiple light signals for superior crop performance

(continued)

Table 2 (continued)

Photoreceptor type	Light spectrum sensitivity	Molecular structure	Signaling pathway components	Biological functions	Environmental interaction	Agricultural implications	Examples	Future research directions
Opsins	Green, Blue, UV	Retinal-binding proteins	Opsin homologs in algae	Photosensory signaling in primitive plants	Light quality, aquatic conditions	Potential bioengineering applications for land plants	Chlamydomonas reinhardtii	Adapting opsin-based signaling to terrestrial crops for novel light-driven processes
Redox-Based Sensors	Multiple spectrums	Reactive oxygen species (ROS) as signaling molecules	ROS sensors, HY5, MAPKs	Stress signaling, light acclimation	High light intensity, oxidative stress	Developing ROS-sensitive pathways for enhanced stress tolerance	Arabidopsis, wheat	Targeting redox signaling to improve light stress management in fluctuating environments
Cyanobacterial Phytochromes	Red, Far-red	Linear tetrapyrrole chromophores	Cph1, PixJ, RcaE	Photosensory adaptation in cyanobacteria	Light quality, aquatic conditions	Utilizing cyanobacterial photoreceptors for synthetic biology in crop improvement	Synechocystis sp.	Translating cyanobacterial light-sensing systems for advanced crop bioengineering
Photoreceptor Networks	All spectrums	Integrated pathways of multiple photoreceptors	PHY-CRY-UVR8 networks, HY5, COP1, SPA	Coordinated developmental and adaptive responses	Light intensity, quality, and spectrum	Comprehensive control of light-driven processes to boost crop resilience and productivity	Arabidopsis, barley	Decoding complex photoreceptor networks for precise agricultural applications

(continued)

Table 2 (continued)

Photoreceptor type	Light spectrum sensitivity	Molecular structure	Signaling pathway components	Biological functions	Environmental interaction	Agricultural implications	Examples	Future research directions
LOV-Kinases	Blue (400–500 nm)	LOV domain linked to kinase activity	Phototropins, FKF1	Light signaling in diverse developmental processes	Light intensity, daylength	Enhancing adaptability of plants to low-light environments	Arabidopsis, pea	Creating synthetic LOV-kinase systems for precise control of light responses
Flavoproteins	Blue, Green, UV	Chromophores: Flavins and pterins	Phototropins, cryptochromes	Multiple roles in light perception and signaling	Light intensity, quality	Broad applications in improving stress tolerance and photosynthesis efficiency	Spinach, Arabidopsis	Expanding knowledge of noncanonical flavoprotein functions in plant signaling
Phytochromes-Cryptochromes Interaction	Red, Far-red, Blue	Combined chromophores of phytochromes and cryptochromes	PHYB-CRY1 complex, HY5	Integrative light signaling for optimized growth	Light quality, photoperiod	Synergistic manipulation for better plant architecture and flowering regulation	Arabidopsis, sunflower	Understanding cross-talk mechanisms to integrate multiple light signals for superior crop performance
Opsins	Green, Blue, UV	Retinal-binding proteins	Opsin homologs in algae	Photosensory signaling in primitive plants	Light quality, aquatic conditions	Potential bioengineering applications for land plants	Chlamydomonas reinhardtii	Adapting opsin-based signaling to terrestrial crops for novel light-driven processes

(continued)

Table 2 (continued)

Photoreceptor type	Light spectrum sensitivity	Molecular structure	Signaling pathway components	Biological functions	Environmental interaction	Agricultural implications	Examples	Future research directions
Redox-Based Sensors	Multiple spectrums	Reactive oxygen species (ROS) as signaling molecules	ROS sensors, HY5, MAPKs	Stress signaling, light acclimation	High light intensity, oxidative stress	Developing ROS-sensitive pathways for enhanced stress tolerance	Arabidopsis, wheat	Targeting redox signaling to improve light stress management in fluctuating environments
Cyanobacterial Phytochromes	Red, Far-red	Linear tetrapyrrole chromophores	Cph1, PixJ, RcaE	Photosensory adaptation in cyanobacteria	Light quality, aquatic conditions	Utilizing cyanobacterial photoreceptors for synthetic biology in crop improvement	Synechocystis sp.	Translating cyanobacterial light-sensing systems for advanced crop bioengineering
Photoreceptor Networks	All spectrums	Integrated pathways of multiple photoreceptors	PHY-CRY-UVR8 networks, HY5, COP1, SPA	Coordinated developmental and adaptive responses	Light intensity, quality, and spectrum	Comprehensive control of light-driven processes to boost crop resilience and productivity	Arabidopsis, barley	Decoding complex photoreceptor networks for precise agricultural applications
Phytochromes-cryptochromes interaction	Red, far-red, blue	Combined chromophores of phytochromes and cryptochromes	PHYB-CRY1 complex, HY5	Integrative light signaling for optimized growth	Light quality, photoperiod	Synergistic manipulation for better plant architecture and flowering regulation	Arabidopsis, sunflower	Understanding cross-talk mechanisms to integrate multiple light signals for superior crop performance

(continued)

Table 2 (continued)

Photoreceptor type	Light spectrum sensitivity	Molecular structure	Signaling pathway components	Biological functions	Environmental interaction	Agricultural implications	Examples	Future research directions
Opsins	Green, blue, UV	Retinal-binding proteins	Opsin homologs in algae	Photosensory signaling in primitive plants	Light quality, aquatic conditions	Potential bioengineering applications for land plants	Chlamydomonas reinhardtii	Adapting opsin-based signaling to terrestrial crops for novel light-driven processes
Redox-based sensors	Multiple spectrums	Reactive oxygen species (ROS) as signaling molecules	ROS sensors, HY5, MAPKs	Stress signaling, light acclimation	High light intensity, oxidative stress	Developing ROS-sensitive pathways for enhanced stress tolerance	Arabidopsis, wheat	Targeting redox signaling to improve light stress management in fluctuating environments
Cyanobacterial phytochromes	Red, far-red	Linear tetrapyrrole chromophores	Cph1, PixJ, RcaE	Photosensory adaptation in cyanobacteria	Light quality, aquatic conditions	Utilizing cyanobacterial photoreceptors for synthetic biology in crop improvement	Synechocystis sp.	Translating cyanobacterial light-sensing systems for advanced crop bioengineering
Photoreceptor networks	All spectrums	Integrated pathways of multiple photoreceptors	PHY-CRY-UVR8 networks, HY5, COP1, SPA	Coordinated developmental and adaptive responses	Light intensity, quality, and spectrum	Comprehensive control of light-driven processes to boost crop resilience and productivity	Arabidopsis, barley	Decoding complex photoreceptor networks for precise agricultural applications

These elaborate light-sensing and signaling networks allow plants to combine external light cues with internal molecular rhythms to fine-tune their physiology to enhance survival and productivity. Together, the dynamics of photoreceptors, recognition of specific wavelengths, and regulation of the circadian clock highlight the intricacy of plant adaptation to fluctuating light landscapes [13, 16, 19].

4 Diurnal Rhythms in Photosynthesis and Carbon Metabolism

Photosynthesis is a vitally crucial biological process modulated at a complex level by the diurnal rhythm, which reflects the axial spin of the Earth and the alternation of day and night. This regulation prevents plants from overinvesting, helping maximize energy capture and partitioning the energy to support growth, development and survival. These environmental cycles occur within the day through light-dependent reactions, the Calvin cycle, and central metabolic processes, which include protein synthesis, starch synthesis, and degradation. They underscore the sophisticated interplay between these processes to respond to the time of day [13, 15, 22].

Photosynthesis starts with light-dependent reactions in the thylakoid membranes of chloroplasts, powered directly by sunlight. Photons from the sun are absorbed by chlorophyll molecules throughout the day, energizing electrons and facilitating the splitting of water molecules (photolysis). It releases oxygen as a waste product and makes ATP and NADPH, the chemical energy reducing the power needed for the following stages of photosynthesis. These products drive the Calvin cycle, which occurs in the stroma of chloroplasts and fixes carbon dioxide into organic molecules. The main enzymes include ribulose–1,5-bisphosphate carboxylase/oxygenase (Rubisco) that promotes the transformation of CO_2 into 3-phosphoglycerate (3-PGA), which is further reduced to triose phosphates and finally used to synthesize carbohydrates.

Light-independent reactions, or the Calvin cycle, named after co-discoverer Melvin Calvin, respond to short and long-term light stimulation. Carbon fixation processes are the most effective when ATP and NADPH can be produced excessively during the day.

Particularly noteworthy is the coordination of starch synthesis and degradation in a diurnal fashion, an essential component of diurnal regulation of photosynthesis. During the light period, plants convert excess photosynthates into transient starch, a storage carbohydrate stored in the chloroplast. This process helps plants cope with the changing availability of carbon resources. Starch synthesis, catalyzed by ATPase, are activated by light and regulated by intermediates like 3-PGA and Pi [12, 14, 24].

Plants break down the stored starch at night when no photosynthesis occurs to support aerobic respiration and growth. There is rigorous control over this degradation to prevent depletion before morning, delivering a steady stream of glucose equivalents throughout the night. The key enzymes α-amylase and glucan water dikinase (GWD) are crucial for converting starch to maltose and glucose, which are then transported to the cytosol to meet the energy and biosynthetic needs. Circadian rhythms coordinate the rates of starch synthesis and degradation with the night length, ensuring optimal use of carbon.

The carbon fixation efficiency is intrinsically greater during the day as the Calvin cycle depends on the products of the light-dependent reactions. This helps Rubisco work as efficiently as possible so that CO_2 can be integrated into their organic molecules with the help of the ATP and NADPH made earlier. Instead, plants possess cues to mitigate carbon flow overnight. One of the most remarkable adaptations is in Crassulacean acid metabolism (CAM) plants, which fix CO_2 at night into malic acid, stored in vacuoles and decarboxylated during the day to release CO_2 for the Calvin cycle. During the day, these plants open their stomata to exchange gases through photosynthesis.

On the other hand, C3 and C4 plants mainly fix carbon during the day. Thanks to the spatial separation of initial CO_2 fixation and the Calvin cycle, C4 plants improve performance under high light and temperature. This minimizes photorespiration, an inefficient pathway that worsens due to Rubisco with its oxygenase activity. Concentrating CO_2 around Rubisco in C4 plants maintains higher photosynthetic efficiency in challenging environments [15, 19, 23].

Plants have an endogenously driven circadian clock that integrates environmental cues to control energy accumulation and use. This clock ties physiological processes together, regulating starch synthesis and degradation and other metabolic activities by predictable day-night cycles. Circadian signals control the expression of genes for enzymes needed for photosynthesis, carbohydrate metabolism, and energy transduction, hence maximizing resource allocation.

Starch degradation timing, for example, is adjusted to match the night length [15]. Shorter nights lead to quicker starch use, and longer nights reduce the degradation rate enough to avoid energy deficits. The regulated fine-tuning of these physiological processes is mediated through clock-associated proteins and signaling pathways, which coordinately modulate enzyme activity and transcription. Circadian rhythms also impact mitochondrial respiration and lead to ATP production and metabolic fluxes according to cellular demand.

Plants can readjust their circadian rhythms in response to environmental oscillations to preserve metabolic homeostasis. This plasticity is essential in stressful conditions such as greying (or nutrient deficiency) when energy-management efficiency is vital for survival. Plants achieve priority resource allocation to growth-related functions, such as repair and defense, under unfavorable conditions by modulating the partitioning of assimilates between growth and storage.

The orchestration of photosynthesis and energy metabolism across the day-night cycle is closely intertwined with developmental events [6, 8, 16]. Light jackets sensed by photoreceptors, including phytochromes and cryptochromes, affect the circadian clock and downstream metabolic pathways. Incorporating photosynthetic activity organizes major points in development (germination, flowering, and fruiting) around available light energy and carbon skeletons.

Overall, this example of diurnal control of photosynthesis demonstrates the interplay between external cues and internal metabolic networks in regulating plant metabolism. By coupling light-dependent reactions, the Calvin cycle, starch metabolism, and circadian inputs, plants fine-tune their energy capture and application to optimize survival and performance in various environments.

5 Stomatal Movement and Water Use Efficiency

Stomatal regulation is central to plant physiology since it mediates gas and water vapor exchange between plants and the environment. Reflex actions and routine processes in plants are carefully controlled by circadian rhythms, which respond to environmental signals. This helps manage water movement in a way that best supports photosynthesis with the right balance of nutrients. Stomatal opening and closing occur according to well-characterized rhythms driven by environmental and internal cues, including light, CO_2 concentration, and circadian clocks.

The stomatal movement showed a regular diurnal behavior pattern where stomata opened in the morning and closed in the evening. During the morning, the stomatal opening is mainly enforced by intensifying light. Light, incredibly blue light, is an essential component since it promotes activation of specific photoreceptors, e.g., phototropins, that trigger a signaling cascade leading to the influx of potassium ions (K +) into guard cells. The accumulation of ions decreases the water potential of guard cells, enabling them to take in water, expand, and open the stomatal pore [1, 5, 21]. Simultaneously, the increase in ambient [CO_2] during daytime morning hours suppresses stomatal opening to some degree, achieving a balance between carbon fixation demand and water conservation.

Additional drivers affect stomatal behavior at various points throughout the day. For example, elevated internal CO_2 concentration due to photosynthesis can modulate the stomatal aperture. Stomata close as light intensity declines during the evening. This is further supported by the loss of guard cell turgor due to K + and another osmolyte efflux. Closing in the evening is essential to prevent water loss through transpiration when photosynthesis stops at sundown.

Light is the most critical external cue for stomatal movements. Blue light directly affects the function of the guard cell, while red light, associated with photosynthetic activity, indirectly changes intercellular CO_2 concentrations. Thus, by balancing CO_2 levels in leaf mesophyll, high levels of CO_2 serve as a signal for partially closed stomata, thereby restricting excessive water loss while allowing sufficient gas exchange to occur during photosynthesis.

Internal circadian clocks are also crucial for stomatal regulation and operate independently of light and CO_2 signaling. These internal biological clocks predict cyclical environmental changes that occur daily and thus enable plants to tune their physiological processes to be optimally suited for their surrounding environmental conditions [17, 18, 23]. For instance, circadian rhythms can prepare guard cells to open stomata in anticipation of sunrise, allowing photosynthesis to begin as soon as light is available. Such a circadian influence guarantees that stomatal behavior will not become entirely reactive but proactively adaptive, aligning physiological responses with probable environmental changes.

Diurnal stomatal behavior is a key regulator of plant transpiration and water use efficiency (WUE). Transpiration, or water loss via stomata, is inextricably linked to stomatal aperture. During the day, this enables CO_2 uptake and photosynthesis but creates a substantial water loss through open stomata. Closure during diurnal periods, including evening and night, reduces water loss, thereby improving WUE. Under water-limiting conditions, precise stomatal control of water loss allows plants to maximize carbon gain without compromising on water conservation.

These mechanisms are complemented in C4 and CAM (Crassulacean acid metabolism) plants by specialized stomatal behaviors that further increase. C4 plants show a highly regulated stomatal response to minimize water loss during high-temperature periods. CAM plants open their stomata for CO_2 fixation at night, leading to very low rates of daytime transpiration. Such adaptations reflect the complexity of stomatal regulation and optimization of water use across different photosynthetic pathways.

Drought-resistant species frequently display specialized adaptations in their stomatal behavior to mitigate drought stress. These plants show reduced stomatal conductance midday (plant losses at discharge) to minimize water depletion [2, 4, 19]. This strategy is aided by quickly reopening stomata in the late afternoon or early evening, when temperature and vapor pressure deficits are lower. This allows the plant to absorb CO_2 with less water loss.

In particular drought-resistant species, circadian regulation is heightened, and so are optimal responses to anticipate and minimize water stress. These species have evolved to have elevated sensitivity to internal and external cues, including soil moisture content and air humidity, facilitating cellular hydration and dehydration avoidance. Desert plants, such as cacti, are an example of CAM photosynthesis, where the diurnal rhythm of stomatal activity is inverted to maximize water conservation.

Furthermore, an osmotic adjustment in guard cells of drought-resistant plants is maintained, allowing stomata to remain functional even at low water potentials. This response includes the accumulation of compatible solutes, such as proline and glycine betaine, which provide turgor in guard cells. These plants utilize biochemical and physiological strategies to endure prolonged droughts, allowing them to perpetuate critical gas exchange processes.

Diurnal regulation of stomatal opening and closure is a convoluted but exquisitely coordinated process involving environmental inputs, intrinsic circadian rhythms, and plant-specific adaptation. This balance between photosynthetic efficiency and water conservation is essential for several species that have evolved to tolerate drought. Further research into these mechanisms will likely yield essential insights regarding possible strategies to help crops adapt to the increased water scarcity wrought by climate change.

6 Hormonal Regulation of Diurnal Rhythms

Within the global symphony of life, plant hormones are the maestro directors conducting the diurnal rhythms of plant life. A complex interplay of physiological processes is orchestrated so that growth, development, and responses to environmental triggers occur in synchrony with the day-night cycle's rhythm. Several hormones (e.g., auxins, abscisic acid (ABA), gibberellins, and cytokinins) exhibit diurnal fluctuations in either concentration or activity, suggesting frequent interactions between the endogenous circadian clocks and key environmental signals such as light and temperature [1, 5, 16, 23]. These dynamic interactions underlie the coordination of growth patterns, stress responses, and resource allocation, highlighting the complexity of plant hormone signaling networks.

Auxins are key in regulating phototropism, gravitropism, and organ development, particularly indole-3-acetic acid (IAA). Indeed, such variations of auxin concentrations have been observed daily in tissues of such as roots and stems. Light-regulated auxin biosynthesis and transport are intricately coupled with the circadian rhythm. This temporal regulation logic allows auxin specification at the subcellular level to map to the best case of potential growth. At the diurnal level, environmental factors can modulate auxin dynamics through photoreceptors like phytochromes and cryptochromes that couple light signals to the circadian clock [1, 17, 21].

Abscisic acid (ABA), a key stress response regulator, has pronounced diurnal oscillations that become enhanced in drought. For example, enzymes like 9-cis-epoxy carotenoid dioxygenase (NCED) are upregulated during high evaporative demand during the day, and ABA biosynthesis is stimulated. This ABA-induced therapeutic production of ABA is paired with stomatal action in a "stomatal behavior" where, hypocritically, stomata close at times of peak transpiration to save water. During the night, ABA levels decrease, and the stomata are open for gas exchange. This diurnal cycling of ABA prevents loss of water from plants and thus mediates cross-talk with circadian clocks and light signaling routes to statically use resources. ABA signaling components, such as PYR/PYL (PYRABACTIN RESISTANCE 1/PYR1-LIKE receptors) receptors and the SnRK2 kinases (SUCROSE NON-FERMENTING 1-RELATED PROTEIN KINASE 2), are also under circadian control, thus demonstrating the integration of the hormone with the global rhythmic circuitry [1, 17, 20, 21]. In addition, ABA modulates drought-responsive gene expression through transcription factors, such as ABA-responsive element (ABRE) binding factors (ABFs), which are expressed diurnally. This coordination highlights the hormone's role in protecting against abiotic stress throughout daily cycles.

Gibberellins (GAs), highly characterized for their role in cell elongation and seed germination, also display diurnal oscillations that govern rhythmic growth. One of the gene families with one of the most well-studied light-dependent cycles is the GA biosynthetic genes, such as GA20ox and GA3ox, that are expressed during the daytime following light cues. This temporal regulation encompasses the GA active window through high photosynthesis efficiency to reach stages where cell growth (or another energy-prompting process, such as shoot elongation) allows. On the other hand, GA signaling is dampened at night to conserve energy and resources [11, 17, 18, 22]. GAs are involved in the cross-talk of other phytohormones (e.g., auxins and cytokinins) to control diurnal growth. For example, they synergistically act with auxins to promote elongation growth, and their interactions with cytokinins contribute to balancing cell division and differentiation.

It is well established that cytokinins (phytohormones involved in cell division and cell differentiation) affect plant growth and development in a diurnal rhythm. The biosynthesis and transport of cytokinins are generally maximized during daytime and regulated by circadian clocks and light signals. Members of the cytokinin-responsive genes, including type-A and type-B response regulators, are expressed in a circadian manner, suggesting that such hormones may be involved in regulating rhythms. Such temporal regulation enables cytokinins to coordinate with other signaling molecules like auxins and GAs to regulate growth and development [4, 16, 22, 23]. For example, we have explained the antagonistic activities of cytokinins and auxins and apical dominance, which modify their antagonistic relationship in response to the illumination and light/dark cycle. In addition, cytokinins coordinate stomatal conductance and nutrient allocation processes, which are naturally rhythmic. This circadian coordination of cytokinins ensures optimal use of resources and physiological equilibrium with daily rhythms. Yes, but there is a critical regulation axis in plant diurnal rhythms: Hormonal cross-talk with circadian clocks and light signaling pathways [6, 17, 18, 22]. The circadian clock comprises interlocked transcriptional feedback loops that control the timing of hormonal biosynthesis and signaling. CIRCADIAN CLOCK-ASSOCIATED 1 (CCA1) and TIMING OF CAB EXPRESSION 1 (TOC1) are among core clock genes that converge with hormonal pathways that fine-tune physiological processes in response to environmental stimuli. CCA1, for example, regulates the expression of auxin-responsive genes, thus linking circadian and growth regulation. TOC1 works similarly to modulate components of signaling pathways involved in stress responses. These photoreceptors coordinate intense light response via communication with circadian clock elements and hormonal modifiers to align growth and development with ambient light conditions. This anatomical setting of plant organisms allows for fine-tuning their resources in the context of changing light and dark conditions; thus, the crosstalk of light and hormones and internal biological clocks represents a typical example of such multilayered regulation of the carboxylation activity for dynamic responses to the light/dark cycle.

Hormones such as auxins, ABA, gibberellins, and cytokinins are essential factors in modulating the repercussions of plants vis-à-vis growth, progress, and stress with diurnal rhythms. Diurnal variation in hormone levels illustrates this dynamic quality of hormonal regulation, driven by circadian clocks and light signaling pathways. The activity of these components in an intertwined fashion creates an environment leading to an optimal adaptation of plants toward stimuli occurring in the environment, along with optimal resource usage and/or mitigating effects of stress, which would otherwise determine plants' morphology, physiology, and developmental features. To provide an example of hormonal integration of stress adaptation in plants, perhaps the most refined is the role of ABA in mediating responses to drought across interlocking diurnal cycles. Moreover, this multifarious relationship emphasizes the importance of studying hormone interplay in the larger context of plant biology and ecological interactions. This could lead to breakthroughs in agricultural methods and plant biochemistry [11, 19, 21, 24].

7 Gene Expression and Diurnal Rhythms

Temporal control of gene expression is a universal feature of biological systems that enables ontogeny to align with environmental cycles. Clock-controlled genes (CCGs) are centrally regulated, and their stable oscillation of expression records the timing of internal circadian clocks. These biological clocks sync biological processes—including growth, stress, and secondary metabolism—to external signals such as light and temperature cycles. Dynamic temporal profiling of CCGs enables organisms to modulate resource allocations, expand survival capacities, and maintain homeostasis under changing environments.

The functions of clock-controlled genes are diverse and include metabolic processes such as cellular division, DNA repair, photosynthesis, and hormone synthesis. Transcriptional-translational feedback loops (TTFLs) are the prototypical driving mechanism that structures rhythmic expression in circadian clocks. The expression of CCGs is typically 24-h periodic, with distinctly defined peaks and troughs in transcriptional activity over diurnal cycles. Early studies demonstrated that this temporality is conserved, e.g., LHY (LATE ELONGATED HYPOCOTYL) and CCA1 (CIRCADIAN CLOCK ASSOCIATED 1) were found to be morning-phased CCG in plants that coordinate expression of photosynthetic genes to sunrise (TOC1 (TIMING OF CAB EXPRESSION 1) is an evening-expressed gene that prepares metabolic pathways for characterizing night processes [1, 2, 18].

At the heart of many rhythmic physiological processes are clock-controlled, including PER (PERIOD), CRY (CRYPTOCHROME), and BMAL1. More specifically, the PER and CRY proteins act as negative regulators of the CLOCK-BMAL1 complex, creating negative feedback loops that facilitate oscillatory gene expression. These mechanisms mediate circadian rhythmic behaviors such as feeding, locomotion, and hormone release. Diurnal fluctuations in oxidative stress and temperature, for example, are supplemented by the expression of circadian CCGs such as HSP70 (HEAT SHOCK PROTEIN 70), which enhances cellular resistance (usually) during the stress response.

At the molecular level, genetic rhythmicity is regulated by a highly intricate matrix of mechanisms involving transcription factors, chromatin remodeling, non-coding RNAs, and post-transcriptional modifications. Transcriptional Control of Biological Oscillations: This is orchestrated by circadian clock components generating transcriptional rhythms. An example of a transcriptionally active couple would be the binding of the CLOCK-BMAL1 heterodimer to the E-box motifs within various target genes' promoters that, in turn, drive their transcription. This activation is prevented by the accumulation of PER and CRY proteins, which negatively regulate

CLOCK-BMAL1 activity and ultimately generate oscillatory expression of some genes [2].

The temporal regulation also has another layer, the post-transcriptional mechanisms, which regulate alternative splicing and, RNA stability and translational control. Specific classes of microRNAs follow circadian rhythms and modulate gene expression at a post-transcriptional level by targeting cytoplasmic mRNA transcripts for degradation and translational inhibition. Further, phosphorylation, ubiquitination, and SUMOylation post-translational modifications regulate the stability and activity of core clock factors, adding extra specificity to temporally regulated transcription [1, 11, 17].

Circadian regulation of growth-associated gene expression ensures that cellular proliferation and elongation events occur at the optimal time of day. In plants, PHYTOCHROME INTERACTING FACTORS (PIFs) facilitate cell elongation at night by interacting with auxin-responsive pathways. Circadian oscillators restrict PIF activities temporally, enabling growth to adjust via photoperiod and temperature adjustments.

Likewise, the TARGET OF RAPAMYCIN (TOR) signaling pathway, a master regulator of cellular growth in all kingdoms of life, integrates circadian and nutrient signaling. This clock-regulated TOR activity aligns tor-mediated anabolism (i.e., protein synthesis and lipid synthesis) and catabolism (i.e., autophagy) with energy demand across the day. GH secretion in animals also has a robust circadian component, with high levels at sleep onset to stimulate tissue repair and growth. One specific group of circadian transcription factors that organizes the time-dependent transcription of signaling components is regulated by genes that mediate signaling through growth hormone (GH) pathways, like IGF-1 (INSULIN-LIKE GROWTH FACTOR 1).

Motor coordination acts through a novel triage program governed by circadian control of stress-responsive genes against external insults. Heat shock proteins, for instance (proteins acting as molecular chaperones), are expressed at rhythmic levels. Their function is to prevent cells from denaturing proteins because of heat. Circadian control of HSP70 expression in plants anticipates protective roles during peak daytime temperatures.

Circadian rhythms of the NRF2 (NUCLEAR FACTOR ERYTHROID 2-RELATED FACTOR 2) pathway are a central mammalian oxidative stress response regulator. Upon entry into the active phase, ROS levels of NRF2 target genes are at their peak to avoid an increase in reactive oxygen species (ROS). Indeed, circadian secretion patterns of stress biomarkers such as cortisol modulate downstream pathways and mediators of stress response, including immune function and metabolic homeostasis [2, 12, 17, 22].

Circadian regulation of the secondary metabolic pathways, such as those for alkaloid, flavonoid, and terpenoid biosynthesis, is also established to maximize resource use and defense. They prevent diseases and maintain the plant life cycle by protecting against UV, resisting pathogens, and adapting to environmental stress due to the biologically significant secondary metabolites such as plant phenylpropanoids (e.g., flavonoids and phenolic acids). Acid azelaic acid plays an active role as a regulatory signal in the immune system of plants by inducing the expression of phenylpropanoid biosynthetic genes, which play the role as an output in plant immune systems to provide defense against pathogens, such as phenylalanine ammonia-lyase (PAL) and chalcone synthase (CHS) which leads to their constitutive expression levels being altered in plants. In the daytime, it aligns with UV light and pathogens; these enzymes will be more active [11, 12, 23, 24].

Microbial biosynthesis of secondary metabolites, including antibiotics and pigments, is under circadian control to align its production with ecological competitive pressures. For instance, genes responsible for metabolizing metabolites like streptomycin in members of the Phylum Streptomyces are temporally regulated and spatially arranged to achieve optimal potency against transient competitors.

The circadian rhythm of these secondary metabolic processes is familiar in animals and influences melatonin production and other neurohormones. Light–dark cycles drive the rhythmic expression of melatonin synthesis genes such as TPH1 (TRYPTOPHAN HYDROXYLASE 1) and AANAT (ARALKYLAMINE N-ACETYLTRANSFERASE) by molecular circadian clocks. These rhythms regulate sleep wake cycles and other behavior patterns relevant to survival.

This temporal regulation of gene expression through clock-controlled genes and rhythmic molecular mechanisms sets up an elementary framework to couple biological processes with environmental cycles. Circadian control adds organismal flexibility, resource utilization, and robustness from growth and stress pathways to secondary metabolism. This knowledge of biological timekeepers also offers insight into the core biology of timekeeping itself. It provides avenues for developing better agricultural practices, chronotherapy strategies for time-of-day disease treatments, and engineering organisms with superior performance in diverse environments. Our understanding of time-based control is evolving, expanding possibilities for biology and biotechnology [12, 17, 20, 22].

8 Diurnal Rhythms and Plant Growth and Development

Diurnal rhythms oscillations that vary over the day based on the cyclic alternations of light and darkness in the environment significantly modulate plant growth and development. These rhythms are controlled by an internal biological clock aligned with the environment's light–dark cycle and influence various physiological processes that promote plant health. The applications of these rhythms to biological processes critical to growth, including cell elongation, leaf growth, flowering, and fruit development, allow plants to coordinate their physiological processes with the external environment. Plant growth is controlled based on a conserved mechanism, but cell elongation is also regulated in a diurnal rhythm [2, 4, 12, 22]. Photosynthetic energy is used in the light phase to build cell synthesis materials. In contrast, in the dark phase, elongation maximizes at the end of that phase, and this kind of expansion results from auxin-induced cell wall loosening. Most elongation processes are also temporally coordinated by a master circadian clock that modulates auxin biosynthetic and signaling pathways. Auxin-responsive genes in Arabidopsis thaliana, for example, show diurnal oscillation and reach a peak twice during the day in response to environmental signals to allow for growth [1, 2, 18]. The left panel depicts how the diurnal regulation of expansion activity, proteins that mediate cell wall loosening, supports the notion that cell elongation is governed by the cross-talk of an endogenous clock and an exogenous light–dark cycle. Diurnal rhythms also regulate another growth process: leaf movement, both nyctinastic (night-time folding) and heliotropic (sun-tracking) behavior. Such movements allow plants to maximize light capture in the daytime but minimize water loss or provide protection to photosynthetic machinery in the nighttime [1, 5, 12, 17]. The circadian clock regulates stomatal movement in leaves, exciting anticipation of light or CO_2 availability. This ultimately modulates the activity of ion channels that mediate stomatal opening in light and stomatal closure in darkness, as well as the well-explained rhythmic expression of clock-regulated genes in stomatal guard cells. This alternating cycle, regulated by circadian rhythms, is a key to effective gas exchange and plant water use. It is a vital factor in growth and productivity across different ecosystems. Flowering is the shift from vegetative to reproductive development, closely associated with diurnal rhythms and photoperiodism. Cullin-PRC, an example of E3 ubiquitin ligase, positions its functions within circadian clock signaling pathways that control the variations in CO, a type of master flowering gene that plays a vital role in flowering under night-length photoperiod conditions. For instance, wheat and barley are long-day flowering plants, and rice and soybean are short-day flowering plants; the former flowering in plants when the day length exceeds some threshold, and the latter flowering in plants when the night length is much longer than the threshold. Instead, the diurnal rhythmic control ensures that the phase of CO expression is correct so that, under the proper photoperiod, this peaks during the day in long-day plants (and peaks at night in short-day plants) to promote flowering [17, 21–23]. Then, chronicle the timing of flower blooms with reproductive fitness to maximize reproductive success. Further, endogenous circadian rhythms regulate fruit's development and ripening via hormone signaling and metabolic pathways tied to these processes. Likewise, the principal ripening hormone, ethylene, exhibits diurnal control in its biosynthetic and signaling cascades. Sugar metabolism displays an analogous circadian regulation and contributes energy and building blocks necessary for fruit development and ripening. In fruiting plants, such as tomatoes, rhythmic expression of cell wall remodeling and pigment biosynthesis genes would facilitate coordinated fruit development and ripening. This temporal system coordination is pivotal in improving fruit's nutritional quality, taste, and commercial value through optimizing ripening and harvest timing. Photoperiodism, the physiological response to the relative lengths of day and night, mediates much of the adaptation of plants to seasonal change. By doing so, plants can synchronize their growth and reproductive cycles with seasonal variations of environmental factors. Since the circadian clock is the primary transducer of photoperiodic responses, these responses are closely linked to our day/night (diurnal) rhythms.

The circadian clock that modulates rhythmic transcription of the genes that encode the photoreceptors (phytochromes responsible for sensitivity to light and cryptochromes sensitive to blue light) mainly determines the timing of the photoperiodic signals. These photoreceptors transmit light signals through downstream pathways that control gene expression, flowering, dormancy, and other seasonal responses [16, 18, 20, 22]. A ratio is also required between temperature and photoperiodic signals, as evidenced by the photoperiodic flowering of temperate plants that occur only following a vernalization process in which prolonged exposure to intense cold becomes effective to induce flower formation. By combining diurnal rhythms with photoperiodism, plants can time their growth and reproduction to the seasonal environment and maximize their chances of survival for themselves and their progenitors. Among perennial species, seasonal adaptation must involve the entrainment of diurnal rhythms to dormancy cycles [21, 23, 24]. Determining Resting State in Different Organisms: A few species of different types of organisms, which include deciduous trees or other long-lived woody species and herbaceous perennials and biennials, are characterized by dormancy induced by short photoperiods and low temperatures, and undergo reduced basal metabolic activities and growth during the winter and are called winter-dormant. The addendum of growth and reproduction in the springtime daily cycles is resynced with environmental input as the days grow longer. This sync is essential for plants to respond to seasonal change. Several responses in light cycles trigger plants, which once again further regulate the growth and development and affect the yield in photoperiod-sensitive crops (rice, wheat, maize, and soybean). Under diverse agro-climatic situations, these replies are paramount to achieving optimum yield.

Photoperiods shorter than the cutoff promote flowering in rice, which is a short-day species. It is because of the expression of the Hd1 gene in the short day (Sd), which accelerates flowering. Conversely, long-day plants, including wheat, exhibit FLOWERING LOCUS and subsequent floral induction upon exposure to prolonged day stimuli. Breeding and genetic engineering-based variants of photoperiodic pathways induction allowed the development of photoperiod-insensitive cultivars that flower and produce grain regardless of the photoperiod. However, photoperiods will die in the growth and flowering processes and other processes such as vegetative growth, seed development, and stress response [1, 5, 16, 21]. The maize, a day-neutral plant, can regulate its growth through leaf elongation and/or increasing the photosynthetic activity of the leaf to develop the leaf to the maximum size possible, independent of the photoperiod. Photoperiod sensitivity is introduced here for soybeans, which helps maximize seed filling and maturity. Therefore, quality and quantity production are derived from this process. Thankfully, the species' genetic and molecular basis of photoperiod sensitivity is better understood through technological advancements in molecular biology and genomics. Genome-wide association studies and quantitative trait locus practitioners identify genes and regulatory pathways controlling the photoperiodic response that guide targeted cross-breeding for plant variety improvement and allow maximizing plant productivity through photoperiodic technologies optimized using remote sensing and phenotyping technologies.

However, it reflects its biological clock photoperiodism, a phenomenon where growth is regulated according to the enduring cycle of day and night. They facilitate extreme performance by tightly controlling the timing of physiological processes, such as cell growth and leaf movement or flowering and fruiting, in a variable environment. Photoperiodism (period of exposure to light) acts as a signal of a seasonal adaptation. It promotes this talent by modulating plants' growth and reproduction activities in response to seasonal changes. These studies of photoperiodic crops provide valuable insights based on the photoperiodic response at the genetic and molecular levels, which help to inform innovative breeding and management of the crops. In the wake of these global challenges, biocontrol of pests and diseases utilizing the efficiency of diurnal rhythm and photoperiodism is one of the most excellent tools available for achieving sustainable agriculture and advancing food security [2, 8, 18, 24].

9 Environmental Factors Modulating Diurnal Rhythms

Biological and ecological systems are heavily influenced by environmental factors such as temperature, humidity, light intensity, and ambient composition. Temperature, for example, influences how enzymes work and affects plants' and animals' metabolic rates and growth. The photosynthesis and respiration of photosynthesizing plants narrow down their operational windows within a thermal range of optimum temperature, and excess temperature raises stress responses, such as the production of heat shock proteins or cold acclimation. Likewise, relative humidity affects transpiration rates, stomatal conductance, and water use efficiency. Transpiration is reduced in high humidity, and conversely, in very low humidity, the plant must face dehydration due to excessive water loss. These confounding factors of TMP and RHis are due to the additional effects of other critical environmental factors, such as soil moisture, wind velocity, and ATm, which can create synergistic or antagonistic interactions [16, 21, 22].

A synergy among these drivers tends to be a non-linear dynamics. So, it can produce more favorable conditions globally for pathogens that thrive on animals and threaten agricultural yields, such as extremes of heat and humidity. On the other hand, dry zones are also marked by high temperatures and low humidity, further resulting in the hazard of evapotranspiration from the perspective of soil moisture, and, as a result, agricultural sustainability deteriorates. In such cases, environmental factors are known to influence the phenology of the organism (for instance, the flowering of plants or the breeding of animals) and thus participate in its functioning by shaping ecosystems. Models of this complexity that combine knowledge of soil, moisture, nutrient availability, and crop performance under various climate scenarios can also help devise more resilient crop varieties.

The circadian clock is an endogenous time-keeping mechanism that orchestrates physiological and behavioral rhythms in response to environmental signals. Large-amplitude temperature oscillations greatly perturb the circadian clocks and define chronology on a cellular scale. How do they accomplish this? For example, plant circadian clocks control essential processes such as stomatal opening, photosynthesis, and hormone production, timing them according to daily temperature cycles. Temperature is additionally conveyed along circadian pathways by thermosensitive transcription factors and post-translational modifications [5, 12, 23].

Another defining property of circadian clocks is temperature compensation, which maintains periodic rhythm stability at varying temperatures. Such a process allows organisms to maintain a circadian rhythm when exposed to variable thermal environments, as described in numerous model organisms. However, extremes in temperature can disrupt circadian rhythms, resulting in physiological impairments. Chronic heat stress, for instance, can misalign what are known as clock-controlled genes, playing a role in both growth and reproduction. This provides an analytical opportunity for improved planting and crop management scheduling based on the relationship between temperature rhythms and circadian clocks in an agronomical aspect. For example, crops modified to exhibit greater circadian robustness may be better adapted to tolerate extremes in temperature. Additionally, temperature-induced circadian disruptions may affect pest dynamics and, thus, pest management, as they may influence crop protection strategies. Light availability finely disentangles many plant diurnal processes, such as photosynthesis, transpiration, and growth [2, 8, 12, 17, 22]. Cloud cover and shade showed substantial changes in diurnal rhythms in light intensity and spectral quality. Light intensity, which is reduced under cloudy conditions, especially at high humidity, reduces rates of photosynthesis and, thus, carbon assimilation and growth. Moreover, reducing the red-to-far-red light atio in the shade can also trigger shade avoidance (SA) responses, including an elongation-growth response and changes in resource allocation.

The growing use of artificial light in human settlements and agricultural landscapes further disturbs these natural diurnal rhythms. Act as a pseudo summer with delayed, aborted flowering and carbohydrate mobilization resulting in very high temperatures. Light pollution interferes with photoperiodic signals and causes a delay in flowering in long-day organisms and an acceleration in short-day plant species. In addition, the spectral quality of artificial light rich in blue or red light can affect different phototropic responses and circadian clock entrainment in various ways.

Artificial light sources such as light-emitting diodes (LEDs) can modify light spectra and photoperiods in agroecosystems, allowing continuous availability for an entire year. However, those systems must be engineered to maximize productivity and minimize energy use and ecological disruption. Understanding how the vegetation, cloud cover, and artificial light interact to alter natural diurnal cycles may help inform sustainable practices like agroforestry, where envisaged shade can create microhabitats that augment biodiversity and ecosystem services.

Climate change, which brings about higher temperatures, changes to rainfall patterns, and more extreme weather events, significantly impacts natural day-and-night cycles. Increased temperatures disrupt the diurnal coupling of processes, e.g., photosynthesis vs. respiration, but they can also dampen or shift ecosystems' metabolic excess. For example, increased nighttime temperatures can elevate respiration rates, decreasing net primary production and carbon sequestration in terrestrial ecosystems.

The increased variability in the precipitation with climate leads to changes in the soil moisture dynamics (or drying up, which changes the plant water usage and the subsequent transpiration cycle within the day). Increased droughts also change stomatal physiology, indirectly impacting photosynthetic efficiency and growth. Phenological shifts in animals' daily activity, such as pollination and foraging, are crucial for plant-pollinator interaction and ecosystem stabilization against climate change.

Acute stress from episodic environmental forces such as heat waves and storms disrupts organisms' diurnal rhythmicity and circadian clock. Thermal stress is driven by increases in regional and/or local high temperatures and directly impacts the performance of enzymes and the photosynthetic apparatus. At the same time, solar storms cause physical damage at a vegetation level and modification of light interception and diurnal patterns. Changes in the timing or intensity of sunlight from variable patterns in cloud cover also affect light-dependent processes that have ripple effects throughout the ecosystem.

Climate change also drives long-term changes in the characteristics of diurnals. Thermal and light regimes change due to organisms' plastic or genetic adaptation. However, such changes may not be sufficient to match rapidly evolving environments with the workings of species interacting. Daisy is the best; for example, as some plants change, their flowering time will not align with when their pollinators are active, disrupting reproduction and reducing biodiversity.

Integrated strategies that combine conservation, sustainable land use, and technology innovations remain the optimal tools for mitigating the impacts of climate change on diurnal cycles. For example, breakthroughs in climate-resilient crop breeding, precision agriculture, and renewable energy could bolster adaptation strategies that will help protect our ecosystems and food in a warming world.

10 Practical Applications in Agriculture

The manipulation of diurnal rhythms to improve crops has relied on understanding the intricate biological processes regulated by circadian timekeeping in plants. Based on proteins, these internal clocks synchronize physiological and developmental processes to respond to environmental cues, including light and temperature [11, 19, 21, 22]. The optimized scheduling of light and temperature provides a beneficial approach for improving crop production and quality in the context of controlled environment agriculture (CEA). In contrast, artificial lighting systems (e.g., light-emitting diodes [LEDs]) enable unprecedented light intensity, spectrum, and duration fine-tuning. There may be possibilities to tune system performance to a plant's natural circadian patterns to enable optimum photosynthetic efficiency and growth rates. Another non-niche example is that know-how has been gained on when to present plant(s) to r/b light spectra, hence increasing, for instance, chlorophyll absorption and flowering, whereby crops like tomatoes and lettuce can thrive through this process. In addition, dynamic light scheduling, where light intensity is increased stepwise, resembles natural sunrise and sunset intensity, reducing photoinhibition and photodamage and further synergizing energy use for CEA systems.

Temperature is equally vital in optimizing circadian rhythm. The difference between daytime and night temperature, known as thermoperiodicity, affects various metabolic processes, including respiration, uptake of nutrients, and the functioning of enzymes. Cooler nighttime temperatures could help save plant energy, while warmer daytime temperatures could improve metabolism and growth. Here, systems managing temperature with precise heating and cooling technologies ensure optimal thermoperiod conservation to enhance the growth cycle and increase yield.

We would need sophisticated sensor systems and automation platforms that automatically adjust the light and temperature schedules mentioned above with our diurnal rhythms. Machine learning is used for dynamic adaptations, while the Internet of Things (IoT) allows for real-time monitoring of plant responses, growth status, and environmental conditions. Having recognized this, the nutrient-regulated crops can grow and thrive when you generate natural cycles as much as they can be available outdoors, enabling you to achieve sustainable, long-term farming. Controlled-environment agriculture (CEA) is transforming and allowing food to be grown in all seasons, even in urban and climate-stressed environments [2, 18, 22, 23]. Adequate artificial lighting regimes are considered key to successful CEA, enabling crops to receive control over ideal and well-distributed light environments independent of everyday weather. LEDs and other advanced lighting technologies allow customized light quality, timing, and intensity; these are powerful tools for optimizing plant performance. Artificial lighting is more than a surrogate for sunlight in CEA applications. Spectral manipulation with LED emission devices enables researchers and agronomists to target plant morphology, phenology, or secondary metabolite production. For example, adding far red light promotes stem elongation and leaf area in leafy greens, which enhances light interception and biomass accumulation. Favorable UV-B light treatments have been shown to increase phenolic compounds, improving nutritional content in crops like kale and spinach.

Artificial lighting regimes also influence the photoperiod (when lights are ON or OFF), another critical factor. Many crops display photoperiodic responsiveness in flowering and reproductive development, requiring specific daylengths. Short-day plants, like soybeans, are coaxed to flower under artificially curtailed daylight cycles; long-day plants, like wheat, flower as the days lengthen. Growers can manipulate certain photoperiods to initiate blooming during precise points, allowing them to harvest on a synchronized schedule and allowing for more output in a single batch.

Light regimes affect crop productivity and the energy efficiency demand issue in CEA. It reduces operational costs while maintaining yield and applying dynamic lighting (e.g., lowering lights when crops are not photosynthesizing or when a driveway sensor is activated) in a split sense to minimize energy. This combination of the best properties of both types of light considerably affects energy efficiency because it is a hybrid of sunlight and artificial lighting. The continued advancement of lighting technology will expand the role of lighting in developing sustainable and productive CEA systems.

So, it is worth noting that gradient sensors have been studied as a potential means for optimizing growth conditions and plant resistance to environmental and biotic stressors, thus improving diurnal rhythmicity. Diurnal patterns regulate key physiological responses that control plant resilience to abiotic and biotic stressors, including drought, heat, and insect attack. So, resilience is increased (while avoiding damaging stress), which allows more consistent productivity working through the different climate periods by harmonizing with these natural diurnal flows [14, 17, 19, 20].

Circadian regulation of stress-responsive genes is prevalent in plants, expressing at optimal times of day to antiviral threats, predisposing plants to subsequent attacks. One example is the diurnal regulation of stomata opening and closure to modulate plant water-use efficiency and drought resistance. By timing irrigation regimes with the times when stomata will be open, plants can be hydrated to a maximum, but will transpire with no excess water. Likewise, diurnal oscillations adjust the levels of ROS and the synthesis of antioxidant enzymes, which are vital for preventing oxidative damage from high-light intensity or abiotic stress.

In addition to abiotic stress tolerance, circadian rhythms regulate the plant defense systems against pathogens and pests. For instance, plants accrue secondary metabolites, including time- and tissue-specific alkaloids and flavonoids that can inhibit herbivory or the growth of microorganisms. Integrating pest management practices (e.g., application of pheromone traps or biocontrol agents) with peak plant defenses could significantly ensure pest suppression at minimal cost and reduce the need for chemical/pesticidal products.

The researcher said bringing this research on the circadian rhythm into crop-breeding programs is an exciting opportunity to make crops more climate-resilient and productive. For example, optimized crops with wild-type circadian clocks have been proven beneficial for photosynthetic efficiency, growth rates, and stress tolerance. These crops' newly tuned circadian clocks link this to altered or wild-type circadian cycles and their impact on stomatal conductance or photosynthetic rate.

Recent advances in genomics and phenomics have also enabled us to identify core genes that mediate these circadian regulations. In contrast, gene editors such as CRISPR-Cas9, among others, can also edit specific genes to regulate the diurnal responses that will lead to the development of certain tailored varieties. For example, changes in the expression of clock-related genes, such as TOC1 and CCA1, have been correlated with improved drought tolerance and increased biomass of model plants. Global food security would have significant implications if these results could be applied to staple crops like rice and maize.

These diurnal rhythms can be exploited using breeding programs to maximize plants' crop yields in controlled conditions. Breeders can prevent this by screening for environmental and nutritional stress tolerance and breeding varieties specifically for CEA systems designed for light and temperature responsiveness. These varieties could show improved light–use efficiency, earliness, or longer shelf life, with agricultural needs [1, 4, 14, 16, 21].

More importantly, the science of diurnal rhythm is to become applicable beyond the conventional breeding frontiers, venturing into more sophisticated phenotyping platforms. This would transfer diurnal traits to a large scale, e.g., leaf movement, photomorphogenesis, and associated growth metrics, which can now be monitored using high-throughput imaging and sensor technologies. So, this information may aid in breeding decision-making and developing climate-resilient, high-yielding crop varieties.

Harnessing diurnal rhythms is a complicated strategy for improving our crops. A valuable configuration for enhancing agricultural productivity and sustainability may be coupling diurnal rhythms from optimized light and temperature schedules in controlled environments to improved stress resilience and climatic adaptation. Continued progress and creative thought in this space will be critical to solving the challenges of producing food for an increasing population in a changing environment.

11 Conclusion

Circadian response to the alternative light and dark cycle is a potential energy source for plant growth, development, and survival. These rhythms enable plants to anticipate environmental changes and help to regulate physiological processes regulated by a complex circadian clock. Several key findings have highlighted the importance of diurnal rhythms to photosynthetic and nutrient uptake, hormone signaling, and stress responses. Photosynthesis, the foundation of all plant productivity, peaks at certain times of day. However, plant physiology is primarily driven by the circadian clock, which controls the expression of genes and the actions of enzymes. Similarly, the opening and closing of stomata, which play a significant role in gas exchange and water use efficiency, have diurnal rigidity, where the light and dark phases are reflected, to maximize carbon dioxide consumption while limiting water loss. Just as energy status governs metabolism, nutrient uptake, for example, nitrogen assimilation, also undergoes diurnal changes to accommodate metabolic requirements in conjunction with energy content. Auxins, cytokinins, abscisic acid (ABA), and many other hormones are tightly controlled in a time-dependent manner by the circadian clock to mediate growth, cell division, and stress responses. The circadian clock coordinates plant immune responses, thereby modulating resistance to biotic stress at specific times, to name a few. Synchronization to an external environmental signal, such as temperature or light intensity, for example, supports homeostasis and, hence, resilience.

Therefore, plant responses to diurnal cycles have important implications for improving agricultural productivity and resilience. Insights gleaned into the circadian regulation of photosynthesis and nutrient use efficiency can be related to potential captures of higher yield and organic resource-use efficiency (O/RE) in crop breeding programs. This led to the hope that the timing of photosynthetic peak activity could be optimized to give biomass production a multiplier effect; genetic manipulation of clock components may help here. Maximizing fertilizer efficiency and minimizing impact on the environment can also be achieved by managing the placement of nutrients according to the plant's uptake rhythms. An improved understanding of the temporal regulation of hormone signaling may guide strategies to enhance stress tolerance during adverse conditions, particularly extreme drought, salinity, and temperature. Exploiting circadian clock modulation of defense responses: A potential avenue for crop improvement. In addition, as climate change brings new shifts in prominent environmental cues like temperature and day length, such crops may maintain higher productivity and resilience under variable conditions due to circadian clocks that are in better sync with ecological cycles. Custom light–dark cycles can also improve growth and resource use efficiency in controlled environment agriculture, e.g., vertical farming, as these cycles can be tailored to the optimal functioning of the circadian clock.

Research on the genetic designing of circadian clocks can lead to better plant sensitivity during diurnal cycles. In recent years, strategies like CRISPR-Cas9 for gene editing have magnified the possibilities of targeting genes and their regulators to achieve phenotypic traits of interest. It will be required to identify and characterize key components of the clock in a broad range of plants to broaden the potential of circadian-derived interventions. Ongoing and future studies into the complex interplay between circadian rhythms and epigenetic modification from external signals may offer further understanding of the mechanisms through which heritable variations in diurnal rhythm responses can facilitate persistent crop improvements across generations. Plasticity, or how circadian clocks are adapted to novel environmental conditions, is another exciting area of research. Establishing how plants recalibrate their clocks with changes in photoperiod, temperature, and other abiotic conditions will be key for this endeavor toward generating climate-tolerant crops. Integrating these quasi-synoptic omics technologies (transcriptomics, proteomics, metabolomics, etc.) within the most recent advances in circadian biology could thus provide an integrative view of the spatial organization of plant physiology. It can then harness modeling, which predicts interactions of circadian networks with the environment to provide predictive crop management approaches. Ultimately, integrating circadian biology with related fields, including agronomy, ecology, and biotechnology, through interdisciplinary research will be essential to bring knowledge from the lab to the field to support sustainable agriculture.

References

1. Anami S, Yamashino T, Kikuchi H, Suzuki R, Aoki S (2024) Nuclear interactions between the Pseudo-Response regulator clock proteins and the multi-step phosphorelay mediator histidine-containing phosphotransfers in the moss physcomitrium patens. Biochem Biophys Res Commun 733:150734. https://doi.org/10.1016/j.bbrc.2024.150734
2. Asar Y, Van Der Wal C, Ho SYWBT (2024) Reference Module in Life Science. Molecular Clock, Elsevier. https://doi.org/10.1016/B978-0-323-95502-7.00050-6

3. Burckard O, Teboul M, Delaunay F, Chaves M (2025) Benchmark for quantitative characterization of circadian clock cycles. BioSystems 247:105363. https://doi.org/10.1016/j.biosystems.2024.105363
4. de los Reyes P, Serrano-Bueno G, Romero-Campero FJ, Gao H, Romero JM, Valverde F (2024) Constans alters the circadian clock in Arabidopsis thaliana. Mol Plant 17(8):1204–1220. https://doi.org/10.1016/j.molp.2024.06.006
5. del Olmo M, Legewie S, Brunner M, Höfer T, Kramer A, Blüthgen N, Herzel H (2024) Network switches and their role in circadian clocks. J Biol Chem 300(5):107220. https://doi.org/10.1016/j.jbc.2024.107220
6. Fu M, Liu S, Che Y, Cui D, Deng Z, Li Y, Zou X, Kong X, Chen G, Zhang M, Liu Y, Wang X, Liu W, Liu D, Geng S, Li A, Mao L (2024) Genome-editing of a circadian clock gene TaPRR95 facilitates wheat peduncle growth and heading date. J Genet Genomics 51(10):1101–1110. https://doi.org/10.1016/j.jgg.2024.05.011
7. Gems D, Virk RS, de Magalhães JP (2024) Epigenetic clocks and programmatic aging. Ageing Res Rev 101:102546. https://doi.org/10.1016/j.arr.2024.102546
8. Hubbard CJ, Harrison JG, McMinn R, Bennett Ponsford JC, Maignien L, Ewers B, Weinig C, Heck M (2023) Rhizosphere microbial community composition shifts diurnally and in response to natural variation in host clock phenotype. MSystems 8(3). https://doi.org/10.1128/msystems.01487-21
9. Ishibashi M, Nakamichi N, Hayashida Y, Kazumori H, Otagaki S, Matsumoto S, Oikawa A, Shiratake K (2024) Genome-wide identification of clock-associated genes and circadian rhythms in Fragaria × ananassa seedlings. Plant Gene 40:100470. https://doi.org/10.1016/j.plgene.2024.100470
10. Kim S-C, Edgeworth KN, Nusinow DA, Wang X (2023) Circadian clock factors regulate the first condensation reaction of fatty acid synthesis in Arabidopsis. Cell Rep 42(12):113483. https://doi.org/10.1016/j.celrep.2023.113483
11. Li J, Qiu J-X, Zeng Q-H, Zhuang Y, Zhang N, Xu S-X, Jin J, Dong Z, Chen L, Huang W (2023) OsTOC1 plays dual roles in the regulation of plant circadian clock by functioning as a direct transcription activator or repressor. Cell Rep 42(7):112765. https://doi.org/10.1016/j.celrep.2023.112765
12. Lian J-W, Li S-Y, Clarke RB, Howell SJ, Meng Q-J (2025) Can we utilise the circadian clock to target cancer stem cells? Cancer Lett 611:217360. https://doi.org/10.1016/j.canlet.2024.217360
13. Makni S, Acket S, Guenin S, Afensiss S, Guellier A, Martins-Noguerol R, Moreno-Perez AJ, Thomasset B, Martinez-Force E, Gutierrez L, Ruelland E, Troncoso-Ponce A (2024) Arabidopsis seeds altered in the circadian clock protein TOC1 are characterized by higher level of linolenic acid. Plant Sci 344:112087. https://doi.org/10.1016/j.plantsci.2024.112087
14. Neto DFM, Araújo WL (2024) Spotlight: orchestrating nitrogen metabolism: histone modifications and its interplay with the biological clock. J Plant Physiol 303:154369. https://doi.org/10.1016/j.jplph.2024.154369
15. Nishikawa S, Murase H, Fukuda H (2013) Theoretical control of the plant biological clock by the periodic dark pulses. IFAC Proc 46(4):211–213. https://doi.org/10.3182/20130327-3-JP-3017.00048
16. Perez-Garcia P, Serrano-Ron L, Moreno-Risueno MA (2022) The nature of the root clock at single cell resolution: principles of communication and similarities with plant and animal pulsatile and circadian mechanisms. Curr Opin Cell Biol 77:102102. https://doi.org/10.1016/j.ceb.2022.102102
17. Reis DJ, Bahraini NH, Yen P, Postolache TT (2023) Body rhythms/biological clocks (Friedman HS, C. H. B. T.-E. of M. H. (Third E. Markey (eds.); pp 257–272). Academic Press. https://doi.org/10.1016/B978-0-323-91497-0.00166-1
18. Shrestha K, Zogli P, Pingault L, Grover S, Cardona JB, Louis J (2024) Disruption of the sorghum circadian clock impacts sorghum-sugarcane aphid interaction dynamics and aphid feeding behavior. Plant Stress 11:100407. https://doi.org/10.1016/j.stress.2024.100407

19. Siqueira JA, Batista-Silva W, Zsögön A, Fernie AR, Araújo WL, Nunes-Nesi A (2023) Plant domestication: setting biological clocks. Trends Plant Sci 28(5):597–608. https://doi.org/10.1016/j.tplants.2023.01.009
20. Sorkin ML, Nusinow DA (2021) Time will tell: intercellular communication in the plant clock. Trends Plant Sci 26(7):706–719. https://doi.org/10.1016/j.tplants.2020.12.009
21. Vo BT, Mas P, Johannes F (2024) Time's up: epigenetic clocks in plants. Curr Opin Plant Biol 81:102602. https://doi.org/10.1016/j.pbi.2024.102602
22. Wang S, Shi K, Xiao Y, Ma W, Hong Y, Feng D, Zhao J (2024) Circadian clock shapes diurnal gene expression patterns linked to glucose metabolic processes in Chinese cabbage. J Integr Agric. https://doi.org/10.1016/j.jia.2024.08.026
23. Xie D, Zhong S, Luo M, Xu J, Zheng R, Luo J, Wang Y, Guo Y, Guo L, Wu B, Lu D (2024) Disruption of local circadian clocks in aristolochic acid-induced nephropathy in mice. Phytomedicine 135:156235. https://doi.org/10.1016/j.phymed.2024.156235
24. Ye C, Micklem CN, Saez T, Das AK, Martins BMC, Locke JCW (2024) The cyanobacterial circadian clock couples to pulsatile processes using pulse amplitude modulation. Curr Biol 34(24):5796–5803.e6. https://doi.org/10.1016/j.cub.2024.10.047
25. Zhu X, Wang H (2024) Revisiting the role and mechanism of ELF3 in circadian clock modulation. Gene 913:148378. https://doi.org/10.1016/j.gene.2024.148378

Chapter 3
Environmental Rhythms and Regulation of Plant Morphogenesis: Temporal Architectures—The Environmental Clocks of Plant Life

Urszula Zajączkowska

1 Introduction

For sessile organisms, temporal information is critical: it enables coordination of growth with predictable changes in light, temperature, and season. Plants rely on internal biological clocks to interpret and respond to these temporal cues. In turn, these clocks affect the adjustment of developmental processes [9]. The rhythms of plant life, especially those synchronised with daily and seasonal cycles, are pivotal to morphogenesis.

Among temporal signals, the daily alternation between light and darkness is the most universal. Through evolutionary adaptation, plants have developed circadian systems that allow them to align internal processes with the solar day. These clocks regulate a wide range of physiological and developmental phenomena, including photosynthesis, hormone production, leaf movement, and flowering time [22, 23]. The ability to anticipate dawn or dusk provides a functional advantage, ensuring that energy-demanding processes occur at optimal times.

This anticipatory regulation extends to developmental timing. Leaf initiation, stem elongation, branching, and reproductive transitions often occur in specific time windows influenced by circadian or seasonal rhythms [28, 37]. The circadian system functions as a temporal gate, enhancing or suppressing developmental signals based on time of day. For example, flowering may be induced only when a promotive light signal coincides with a sensitive phase of the circadian cycle [30].

Photoperiodism—the response to day length—is a clear demonstration of how environmental rhythms influence morphogenesis. Many species use photoperiod as a seasonal signal to initiate flowering, dormancy, or other irreversible changes in

U. Zajączkowska (✉)
Department of Forest Botany, Warsaw University of Life Sciences, Warsaw, Poland
e-mail: urszula_zajaczkowska@sggw.edu.pl

A. Kothari Chhajer and V. Singh Gour (eds.), *Introduction to Rhythms in Plants*, Studies in Rhythm Engineering, https://doi.org/10.1007/978-981-95-3592-7_3

development. This is particularly important in annual plants that must complete their life cycles within favorable conditions. Short-day and long-day plants differ not only in their response thresholds but also in the temporal regulation of underlying developmental pathways [8, 38].

Temperature is another important rhythmic cue. Daily fluctuations and seasonal patterns influence germination, flowering, and dormancy. The integration of thermal and photoperiodic information refines developmental decisions, particularly under variable field conditions. Some plants require cumulative thermal exposure (vernalization) before initiating certain transitions, indicating a capacity to measure time not only through cycles but also through accumulation [10, 24]. This temporal regulation leaves structural and anatomical traces. For example, seasonal rhythms in cambial activity determine the patterning of xylem and phloem differentiation. Annual rings in woody plants are not merely records of environmental conditions, but also of rhythmic control over cell division and expansion [18, 34].

Similarly, rhythmic growth patterns in herbaceous plants can produce regular internode spacing and leaf arrangement, reflecting internal timing mechanisms shaped by external cues.

Developmental plasticity depends on the synchronization of internal rhythms with environmental cycles. Disruption of this synchrony—for example, through artificial lighting or climate anomalies—can reduce growth efficiency and affect key developmental transitions [17, 21]. Field studies have shown that gene expression patterns differ markedly under natural conditions compared to controlled environments, emphasizing the importance of studying plants within the full context of environmental rhythmicity [20]. Different species—and even different genotypes within a species—show variation in their sensitivity to environmental timing. This variation has been utilized in plant breeding to adapt crops to different latitudes and growing seasons. However, the selection for fixed flowering times or reduced dormancy can reduce flexibility, making modern cultivars more susceptible to seasonal anomalies [5, 29]. Current changes in global climate patterns further highlight the importance of understanding the temporal regulation of plant development. Altered temperature regimes, unpredictable rainfall, and disrupted photoperiods—due to factors such as light pollution or urban heat islands—may decouple internal clocks from external conditions. This desynchronization could impair growth and reproductive success, particularly in species with narrow developmental windows or rigid timing requirements [7, 31].

In sum, the temporal organization of plant development is an essential but often underappreciated factor in morphology. Internal rhythms allow plants to align form-building processes with environmental opportunity and constraint. Rhythmicity in plant biology is not a secondary feature; it is a primary condition under which growth and differentiation occur. Morphogenesis proceeds not in a continuous stream, but in phases defined by time, regulated by both endogenous clocks and environmental cycles.

2 Rhythms of Cambial Activity and the Periodicity of Vascular Development

The vascular cambium is one of the most rhythmically active tissues in plants. Responsible for producing xylem and phloem during secondary growth, it operates in close coordination with environmental cycles. Yet unlike primary meristems, which are developmentally continuous, cambial activity is intermittent, subject to clear periods of dormancy and reactivation. This periodicity is not incidental. It reflects a deep integration between developmental potential and environmental time—particularly the cycles of light, temperature, and water availability that define plant seasons. In temperate species, the annual cycle of cambial activity is especially pronounced. Cell division begins in spring, accelerates in early summer, and slows or halts in late summer and autumn. These transitions correlate strongly with rising and falling temperatures, but also with changes in daylength. Bud burst and leaf expansion tend to precede cambial reactivation, suggesting that signal integration across multiple organs determines the timing of vascular growth [19, 25]. Trees do not resume cambial activity solely because temperatures rise, but because a particular combination of cues—accumulated chill, sufficient light, internal carbohydrate levels—align to unlock a new growth phase. This seasonal periodicity has anatomical consequences. Early-season xylem is composed of large, thin-walled cells adapted for efficient water transport. Later, as growth slows and water availability becomes less predictable, the cambium shifts toward producing denser tissue with smaller cells contributing to mechanical strength and embolism resistance. The pattern is not uniform across species or years, but the rhythm is unmistakable. Tree rings are physical records of this cyclical developmental pattern, with each band encoding the temporal structure of cambial activity during a particular year [6, 35].

In tropical regions, where photoperiod is less variable and frost absent, the patterns are more diverse. Some trees exhibit continuous cambial activity throughout the year, while others display two or more growth flushes separated by short pauses. Moisture becomes a dominant factor in these ecosystems. In species like Grewia tiliaefolia or Mimusops elengi, vascular differentiation follows seasonal rainfall, and growth pauses coincide with drought or reduced transpiration [11, 14]. Even in evergreens, vascular development follows rhythm, though the cues are more subtle than in deciduous temperate taxa. What unites both tropical and temperate species is not the identity of the cue but the regularity of the pattern. The cambium is a developmental gatekeeper, and its cycles reflect environmental windows when growth is not only possible, but also advantageous. Starting too early may expose new xylem to freeze–thaw damage, continuing too late may compromise reserve storage. Rhythmicity thus protects the plant from temporal mismatch—a failure to align structure with season.

Daily rhythms also play a role. In certain species, microscopic studies have shown changes in cambial turgor and wall softening that vary across the day–night cycle. While not as extensively studied as seasonal rhythms, these fluctuations suggest that even within a growing season, the cambium does not act continuously but in phase with diel variation in water availability and assimilate flow [33]. Such fine-scale rhythmicity likely contributes to variation in cell size and wall thickness even within a single ring. The physiological basis of these rhythms is complex but increasingly understood as involving not only temperature and light, but also internal signaling related to hormonal balance and carbon availability. Auxin gradients influence the positioning and differentiation of cambial derivatives, but their effect is gated by environmental rhythms. Hormonal activity without seasonal context is insufficient to sustain cambial division [16]. The cambium is not only hormonally controlled, but rhythmically enabled. These periodic dynamics are not only biological curiosities but ecologically significant. In a changing climate, with increased variability in temperature and precipitation, the timing of cambial activity may become desynchronized from optimal conditions. Early activation followed by late frost can damage new tissues, delayed activation may shorten the productive season. Recent observations suggest that climate anomalies are already disrupting the regularity of growth rings in some regions, creating ambiguous boundaries and uneven vessel structure [26, 27]. Understanding these rhythms, and their flexibility, is critical not only for basic plant science but for applied fields such as forestry and climate adaptation. Trees grow in time, and the cambium—through its cycles of dormancy and activity—is one of the primary mechanisms through which temporal structure is translated into anatomical form.

3 Gravitational and Lunisolar Rhythms in Plant Morphogenesis

The growth and development of plants are closely tied to cyclic environmental forces, most prominently light and temperature. Yet beyond these well-established entrainment cues, evidence has emerged suggesting that subtler forces—particularly those derived from the gravitational dynamics of the Earth-Moon-Sun system—may also play a role in modulating aspects of plant morphogenesis. Although still a developing area of inquiry, studies over the past several decades point to the possibility that lunisolar cycles influence plant physiology at levels ranging from root elongation to stem diameter fluctuations and possibly cambial behavior. Early interest in this domain came from observations of rhythmic leaf movements and stem diameter variations that could not be fully explained by known circadian or photoperiodic mechanisms. In some species, rhythmic fluctuations in stem girth were observed to align more closely with lunar phases than with solar time. Barlow and colleagues reported that the diameter of tree stems oscillates with a periodicity approximating the lunar tidal cycle, suggesting a physiological response to changes in gravitational vector strength or direction [4, 12]. Although the magnitude of these fluctuations is small—on the order of micrometers—the consistency across species and locations raised questions about the underlying mechanisms.

More targeted experimental work has provided tentative support for these hypotheses. In controlled conditions, peppermint (Mentha $\times$ piperita) exhibited growth rhythms in stem elongation and leaf nutation that remained in phase with lunar gravitational cycles, even in the absence of photoperiodic or thermal cues. The authors proposed that tidal acceleration—resulting from the Moon's orbit and Earth's rotation—may act as an entraining signal for certain rhythmic growth patterns [36]. Roots, too, have been investigated in this context. Arabidopsis root elongation rates have shown fluctuations that align with circatidal periods, prompting speculation that root statoliths—gravity-sensing organelles in root cap cells—may respond not only to constant vector gravity but also to minute gravitational changes over time [13]. If true, this would suggest that the plant gravitropic system is not merely responsive to direction, but potentially to time-dependent variation in gravitational intensity. A related line of research has explored electrical potentials and bioelectrical signaling in the cambial zone, where minute oscillations in voltage appear to follow external tidal cycles. Electrodes inserted into the phloem and cambium have recorded rhythmic electrical activity that, in some cases, varies with lunar position [1]. While speculative, such findings suggest that the perception of lunisolar rhythms may not be purely mechanical, but involve complex physiological integration. Skepticism remains justified. Many of the observed patterns could result from uncontrolled environmental variation, statistical coincidence, or unaccounted correlations with other cycles such as air pressure or soil water potential. Moreover, the physical force associated with lunar tidal acceleration is exceedingly weak—several orders of magnitude smaller than those commonly required to trigger known plant responses. Yet, as Fisahn and Barlow point out, biological systems often operate at thresholds and through amplification mechanisms that allow them to respond to minimal inputs if those inputs are periodic and persistent [3].

The idea that plant development might be entrained, in part, by lunisolar cycles is not new. In earlier ecological literature—and even in traditional agricultural calendars—references to lunar timing and "sap flow peaks" around full moons are common, though rarely quantified. With the advent of precise measurement tools such as kymographs, dendrometers, and electrochemical sensors, the biological plausibility of these cycles can now be tested with greater resolution [2]. Still, the question remains open as to whether such rhythms play a structural role in morphogenesis, or whether they are secondary modulations layered atop stronger, well-characterized environmental regulators. For now, it appears that lunisolar forces may subtly influence the timing and coordination of ongoing developmental processes—perhaps by modulating tissue hydration, cell wall elasticity, or internal transport—but do not override the primary circadian or seasonal control systems. In this light, their role may be likened not to an architectural blueprint, but to a temporal fine-tuning—a way of maintaining synchrony and coherence in systems already primed by stronger stimuli.

These ideas are not without controversy. The broader field of plant neurobiology, within which some of this research is situated, has been criticized for overextending metaphor and underdefining mechanisms. Yet there is value in exploring rhythmic phenomena beyond the dominant paradigms, especially when observations remain consistent across systems and conditions. As with early circadian research, skepticism must coexist with openness to patterns not yet fully understood. Whether lunisolar rhythms will ultimately be recognized as functional regulators of plant development or remain physiological curiosities is uncertain. But their study invites a broader reflection on what counts as an environmental cue and how even the most minute periodic forces might become part of a living organism's developmental language.

4 Temporal Coordination and Environmental Adaptation in Plant Morphogenesis

Plant development unfolds within a temporal landscape shaped by rhythmic environmental forces. These rhythms—diurnal, seasonal, and even lunisolar—do not act in isolation. They intersect and interact, structuring both the pace and form of morphogenesis. While much attention has been given to individual rhythms, their integration represents a deeper layer of biological organization—one in which plants not only respond to time, but organize their bodies through it. Daily rhythms govern short-term physiological cycles: stomatal opening, photosynthetic efficiency, and hormone oscillations. Seasonal rhythms, by contrast, set the broader framework of developmental phases—when to initiate growth, when to flower, when to enter dormancy. Yet these temporal domains must be coordinated. A plant that begins cell expansion at the wrong time of day may suffer desiccation; one that initiates flowering out of season may miss pollination entirely. Morphogenesis, in this light, is not simply a response to individual cues, but the product of a temporal logic that reconciles multiple periodic signals [8, 22]. This logic is most evident in phenology—the timing of life-history events. Phenological phases such as germination, leaf emergence, and reproduction are not continuous gradients but discrete transitions, often gated by combinations of light, temperature, and internal readiness. These gates function as developmental thresholds, requiring not only the presence of a cue but its coincidence with a specific phase of internal time [15, 23]. This logic reflects earlier insights into the concept of a "*window of opportunity*", a developmental phase during which the plant is competent to respond to environmental stimuli.

As [32] emphasized, morphogenetic responses are possible only when external factors coincide with internal states of readiness. Without this temporal alignment, even strong environmental signals may fail to induce change. Thus, plants operate not merely by reacting to cues, but by filtering them through their internal rhythms and physiological phase. For instance, flowering may occur only when the inductive photoperiod aligns with a permissive phase in the internal seasonal program. Coordination across temporal scales is also observed in cambial activity. Daily variations in turgor and assimilate supply modulate cell division within the cambium, but only during periods of seasonal permissiveness. Cambial dormancy cannot be reversed by transient warmth or moisture if photoperiodic conditions are inadequate. Conversely, the onset of growth may be delayed if thermal thresholds are unmet, despite lengthening days. These interactions reflect a system that integrates multiple environmental dimensions into a coherent developmental response [33, 34]. The complexity of temporal coordination raises the question of plasticity versus constraint. On one hand, plants exhibit remarkable flexibility in adjusting developmental timing to local conditions—a trait evident in ecotypic variation in flowering time, dormancy release, and seasonal leaf drop [24, 30]. On the other hand, certain thresholds remain rigid. Some perennials, for example, require a minimum chilling period before reinitiating growth, regardless of subsequent warmth. In this sense, temporal plasticity is not limitless. It is shaped by evolutionary trade-offs between responsiveness and reliability—between adapting to short-term variability and maintaining alignment with long-term patterns [5].

Integration also plays a role in resilience. Plants that synchronize development with stable cues—such as photoperiod—tend to show more predictable phenologies year to year. In contrast, those relying more heavily on temperature may exhibit greater interannual variability, particularly under climate change. As warming trends decouple historical relationships between temperature and season, the internal logic of morphogenesis may become destabilized [10, 31].

Perhaps the most intriguing dimension of temporal coordination is its possible extension beyond the solar cycle. As discussed in previous chapters, certain studies have identified patterns in root elongation, stem diameter, and electrical signaling that appear to align with lunar or tidal cycles. While controversial, these observations suggest that even extremely weak and slow-varying forces may be integrated into the plant's temporal architecture—not as primary drivers, but as modulatory signals [3, 36]. Their inclusion points to the breadth of temporal sensitivity in plants and to the possibility that development is not only governed by strong cues, but finely tuned by subtle ones. In integrating multiple rhythms, plants do not merely navigate their environments—they construct a temporal identity. This identity is expressed through phenology, architecture, and growth dynamics. It enables ecological adaptation across climates and latitudes, and it is shaped by the constraints and opportunities of rhythmic structure. As such, the study of temporal coordination in plant development is not only a matter of physiology, but of evolutionary and ecological strategy. Morphogenesis unfolds not simply in space, but in time and understanding how plants read and reconcile time is essential to understanding how they grow.

References

1. Barlow PW (2012) The rhythms of trees: the significance of periodicity for development. In: Mancuso S, Baluska F (eds) Long-distance systemic signalling and communication in plants. Springer, Berlin. https://doi.org/10.1007/978-3-642-29110-4_10
2. Barlow PW (2015) Lifting the veil—lunar influences on plant growth and development. Annals Botany 116(2):149–151. https://academic.oup.com/aob/article-abstract/116/2/149/180655
3. Barlow PW, Fisahn J (2012) Lunisolar tidal force and the growth of plant roots, and some other of its effects on plant movements. Annals Botany 110(2):301–318. https://academic.oup.com/aob/article-abstract/110/2/301/2769204

4. Barlow PW, Mikulecky DC, Stoyanov MA (2010) Tree stem diameter fluctuates with the lunar tides. Protoplasma 246(1):25–32. https://doi.org/10.1007/s00709-010-0136-6
5. Bhattacharya A, Khanale V, Char B (2017) Plant circadian rhythm in stress signaling. J Plant Biochem Biotechn. https://doi.org/10.1007/s40502-017-0299-7
6. Bosio F, Oliveira JM, Roig FA, Maria FJ (2016) Seasonal cambial activity and xylem formation in tropical trees from flood-prone forests in Central Brazil. Trees. https://doi.org/10.1007/s00468-016-1383-8
7. Cai Z, Zhang Y, Tang W, Chen X, Lin C, Liu Y (2022) Lux Arrhythmo interacts with ELF3a and ELF4a to coordinate vegetative growth and photoperiodic flowering in rice. Front Plant Sci. https://doi.org/10.3389/fpls.2022.853042/full
8. Cao S, Luo X, Xu D, Tian X, Song J, Xia X (2021) Genetic architecture underlying light and temperature mediated flowering in Arabidopsis, rice, and temperate cereals. New Phytol. https://doi.org/10.1111/nph.17276
9. de Montaigu A, Tóth R, Coupland G (2010) Plant development goes like clockwork. Trends Genet 26(7):296–306. https://doi.org/10.1016/j.tig.2010.04.003
10. Desai J (2019) Evaluating the effects of increasing nighttime temperature on the circadian clock using transcriptional networks. Doctoral dissertation. https://search.proquest.com/openview/ca696f5897af470e394a17b128546e40/1
11. Deshpande BJ, Rajendrababu KC (1985) Periodicity in the vascular cambium of Grewia tiliaefolia Vahl. Ann Bot. https://academic.oup.com/aob/article-abstract/56/1/61/110903
12. Fisahn J (2018) Circadian and tidal oscillations of water potential and root growth in Arabidopsis thaliana. Ann Bot 122(5):735–743. https://academic.oup.com/aob/article-abstract/122/5/735/4823029
13. Fisahn J, Koscielniak J, Barlow PW (2018) Arabidopsis root elongation exhibits rhythmic behavior that aligns with tidal acceleration cycles. Ann Bot 122(5):725–735. https://academic.oup.com/aob/article-abstract/122/5/725/4718565
14. Ghouse AKM, Hashmi S (1983) Seasonal changes in cambial activity and wood formation in Mimusops elengi Linn. IAWA Bull. https://www.sciencedirect.com/science/article/pii/S036725301732025X
15. Gil KE, Park CM (2019) Thermal adaptation and plasticity of the plant circadian clock. New Phytol. https://doi.org/10.1111/nph.15518
16. Gurumurti K, Lal P, Mohan Ram HY (1985) Hormonal regulation of cambial activity and wood formation. In: Recent advances in plant physiology. https://doi.org/10.1007/978-94-009-5139-6_12
17. Iwamoto M, Higo K, Takano M (2009) Circadian clock–and phytochrome–regulated Dof–like gene, Rdd1, is associated with grain size in rice. Plant, Cell Environ. https://doi.org/10.1111/j.1365-3040.2009.01954.x
18. Kim K (2019) Molecular mechanisms for thermostable circadian clock in Arabidopsis and thigmomorphogenetic responses to environmental changes in Brachypodium. Seoul National University Repository. https://s-space.snu.ac.kr/handle/10371/162458
19. Liphschitz N, Lev-Yadun S, Waisel Y (1981) The annual rhythm of activity of the cambium and phellogen in Cupressus sempervirens L. Annals of Botany. https://academic.oup.com/aob/article-abstract/47/4/485/175962
20. Matsuzaki J, Kawahara Y, Izawa T (2015) Punctual transcriptional regulation by the rice circadian clock under fluctuating field conditions. Plant Cell. https://academic.oup.com/plcell/article-abstract/27/3/633/6118139
21. Nomoto Y, Kubozono S, Yamashino T (2012) Circadian clock-and PIF4-controlled plant growth: a coincidence mechanism integrates hormone signaling into photoperiodic control. Plant Cell Physiol. https://academic.oup.com/pcp/article-abstract/53/11/1950/1865027
22. Oravec MW, Greenham K (2022) The adaptive nature of the plant circadian clock in natural environments. Plant Physiol. https://academic.oup.com/plphys/article-abstract/190/2/968/6650635
23. Paajanen P, de Barros Dantas LL, Dodd AN (2021) Layers of crosstalk between circadian regulation and environmental signalling in plants. Curr Biol. https://www.cell.com/current-biology/fulltext/S0960-9822(21)00385-7

24. Patnaik A, Alavilli H, Rath J, Panigrahi KCS (2022) Variations in circadian clock organization & function: a journey from ancient to recent. Planta. https://doi.org/10.1007/s00425-022-04002-1
25. Priestley JH (1930) The seasonal activity of the cambium. New Phytol. https://www.jstor.org/stable/2428433
26. Pumijumnong N, Sathira K, Palakit K (2023) Cambial activity in tropical tree species in response to environmental variation. Forests. https://www.mdpi.com/1999-4907/14/5/1025/pdf
27. Rao RV, Rajput KS (2001) Annual ring formation in some Indian tropical trees. Ann For Sci. https://www.afs-journal.org/articles/forest/pdf/2001/06/35-00.pdf
28. Srivastava D, Shamim M, Kumar M, Mishra A (2019) Role of circadian rhythm in plant system: an update from development to stress response. Plant Sci. https://www.sciencedirect.com/science/article/pii/S0098847218315661
29. Verma S, Negi NP, Narwal P, Kumari P (2022) Calcium signaling in coordinating plant development, circadian oscillations and environmental stress responses. Plant Sci. https://www.sciencedirect.com/science/article/pii/S0098847222001575
30. Wang F, Han T, Chen ZJ (2024) Circadian and photoperiodic regulation of the vegetative to reproductive transition in plants. Commun Biol. https://www.nature.com/articles/s42003-024-06275-6
31. Wang Y, Wu F, Lin Q, Sheng P, Wu Z, Jin X (2023) A regulatory loop establishes the link between the circadian clock and abscisic acid signaling in rice. Plant Physiol. https://academic.oup.com/plphys/article-abstract/191/3/1857/6885543
32. Wardlaw IF (1963) Organization and regulation of growth. vol. 1: the chemical control of growth, McGraw-Hill, New York
33. Wardlaw IF (1990) The control of carbon partitioning in plants. New Phytol. https://doi.org/10.1111/j.1469-8137.1990.tb00524.x
34. Yamamoto Y, Sato E, Shimizu T (2003) Cambial activity and seasonal rhythm in temperate forest trees. Plant Cell Physiol. https://academic.oup.com/pcp/article-abstract/44/11/1119/1939831
35. Yamamoto Y, Sato E, Shimizu T (2003) Comparative genetic studies on the APRR5 and APRR7 genes in control of flowering time and circadian rhythm. Plant Cell Physiol. https://academic.oup.com/pcp/article-abstract/44/11/1119/1939831
36. Zajączkowska U, Barlow PW (2017) The effect of lunisolar tidal acceleration on stem elongation growth, nutations and leaf movements in peppermint (Mentha piperita L.). Plant Bio 19(5):780–788. https://doi.org/10.1111/plb.12561
37. Zhang Y, Ma Y, Zhang H, Xu J, Gao X (2023) Environmental factors coordinate circadian clock function and rhythm to regulate plant development. Cell Cycle. https://doi.org/10.1080/15592324.2023.2231202
38. Zhao J, Huang X, Ouyang X, Chen W, Du A, Zhu L (2012) OsELF3-1, an ortholog of Arabidopsis EARLY FLOWERING 3, regulates rice circadian rhythm and photoperiodic flowering. PLOS ONE. https://doi.org/10.1371/journal.pone.0043705

Chapter 4
Circadian Regulation of Rhizomicrobiome

Reema Kumar and Aditi Kothari Chhajer

1 Introduction

The rhizosphere is the nutrient-rich and interactive soil region surrounding the plant roots where intricate communication occurs between the plant and the microorganisms. It is a complex ecosystem housing a symbiotic plant–microbe holobiont [18]. The microbial community inhabiting the rhizosphere is termed as rhizomicrobiome that encompasses bacteria, fungi, protists, nematodes, and invertebrates [43]. Plants respond to varying environmental signals and carry out nutrient exchange between the roots and the soil, based on which there is a strong selection observed by plants for their rhizomicrobiome [30]. The dynamic and complex network of signaling between the plants and the rhizomicrobiome facilitates both the host and microbial growth and development. Other functional roles of the rhizomicrobiome include biofilm formation to augment nutrient uptake and resilience toward biotic and abiotic stresses [23].

The intricate network between the host plant and the rhizomicrobiome is intensely impacted by the differences in light and dark cycles and ambient temperatures. The circadian rhythm or the internal timing mechanism of plants is widely studied where they are known to maintain a daily rhythm, aligning their internal physiological processes to the external cues from the presence and absence of daylight [7]. This ensures that the biological processes occur during the most pertinent time durations throughout the day, oriented by environmental cues or zeitgeber. In plants, the circadian rhythm regulated by light influences various metabolic pathways such as photosynthesis, chlorophyll synthesis, electron transport, and nitrogen assimilation

R. Kumar (✉)
Lassonde School of Engineering, York University, Toronto, Canada
e-mail: ireema15@gmail.com

A. K. Chhajer
Department of Botany, Sri Venkateswara College, Delhi University, Delhi, India

A. Kothari Chhajer and V. Singh Gour (eds.), *Introduction to Rhythms in Plants*, Studies in Rhythm Engineering, https://doi.org/10.1007/978-981-95-3592-7_4

[11]. Such responses in plants to the diurnal changes in light further create fluctuations in rhizomicrobiome activity. Likewise, the circadian rhythms of microbial communities in the rhizosphere can modulate the flowering time in plants, regulate nutrient availability in the soil, and affect energy metabolism in the plants [9, 27, 48].

This chapter provides an overview of circadian rhythms in plants and rhizomicrobiome and their interactive regulation.

2 Overview of Circadian Rhythms in Plants

Nearly all living organisms, including plants, possess an internal timekeeping system called the circadian clock. This clock allows organisms to predict and coordinate their internal biological functions with the daily environmental cycles [32]. In plants, this central clock is a sophisticated network of genes that activate and suppress each other, forming interconnected feedback loops. These clock genes exhibit specific expression patterns throughout the day. Beyond regulating each other, they also govern numerous physiological processes. These processes, influenced by the clock, may display rhythmic activity, peaking at specific time of the day. Furthermore, the clock can "gate" these processes, making them more sensitive to environmental cues at specific times [13]. This ensures that a plant's responsiveness to light is highest during daylight, to growth hormones at night, and to environmental stressors when adverse conditions are most probable. To effectively tailor responses to external factors, the clock integrates with light and temperature signaling pathways, ensuring alignment between internal and external rhythms [3, 6].

The basic structure of this gene-based oscillator appears to be consistent across various plant species. The importance of the circadian clock is highlighted by the role of these core clock genes in domestication and crop enhancement. Research on *Arabidopsis* reveals that the circadian clock consists of multiple interacting regulatory circuits. Each gene responds to unique environmental signals, and its protein product controls a specific set of outputs, including the activity of other clock components. These components work together to enable responses to diverse environmental inputs and to regulate most of the plant's biochemical and physiological processes. Numerous reviews detail the specific components of the *Arabidopsis* circadian clock and the roles of transcription, splicing, translation, and post-transcriptional signaling in regulating the central oscillator [31, 35].

2.1 Molecular Basis of Plant Circadian Rhythms

The plant circadian clock in *Arabidopsis thaliana* is orchestrated by a complex network of interconnected feedback loops, primarily involving repressor and activator transcription factors whose protein levels fluctuate throughout the day. This central oscillator is structured around three main loops: morning, central, and evening. The central loop features the morning-expressed MYB-like transcription factors Circadian Clock Associated I (CCA1) and Late Elongated Hypocotyl (LHY), which repress the afternoon-expressed Pseudo-Response Regulator (PRR) genes, including Timing of Cab Expression1 (TOC1), PRR5, PRR7, and PRR9. Conversely, TOC1 and other PRR proteins repress CCA1 and LHY, completing a core feedback cycle. CCA1 and LHY, acting as transcriptional repressors, target the evening element (EE) cis-motif found in various clock genes, including the PRRs, and also regulate the evening complex genes Early Flowering3 (ELF3), ELF4, and Lux Arrhythmo (LUX), which are expressed at night and further repress morning and afternoon genes [14, 21, 25].

Beyond these negative feedback loops, the circadian network also incorporates positive regulation. Midday-expressed MYB-like transcription factors Reveille4 (RVE4), RVE6, and RVE8 activate several clock genes, including TOC1, the PRRs, and the evening complex genes. RVE8, in conjunction with Night Light-Inducible and Clock Regulated (LNK1) and LNK2, directly promotes TOC1 and PRR5 expression. Notably, RVE proteins are themselves subject to repression by TOC1 and the PRRs, forming additional negative feedback loops. The antagonistic interplay between CCA1/LHY repressors and RVE activators, both binding to the same EE sites, highlights the importance of the balance of these factors in determining circadian period [15, 31, 36, 45].

Specifically, within the morning loop, CCA1 and LHY repress PRR7 and PRR9, while in the evening loop, TOC1, along with GI, represses an unknown factor that activates TOC1. Furthermore, Gigantea (GI) interacts with Zeitlupe (ZTL) protein to degrade TOC1. The evening complex, comprising LUX, ELF3, and ELF4, forms further interlocked negative feedback loops. Positive regulators, including RVE genes, LWD1/2, and LNK genes, also contribute to the oscillator's dynamics [38].

The circadian clock's output pathways influence a broad spectrum of physiological processes, including reproductive development, hormone production, defense responses, and overall gene expression Fig. 1. The clock also modulates stress acclimatization, hormone signaling, morphogenesis, and carbon metabolism [39]. Circadian gating, the process by which the clock aligns plant responses with environmental cues, is crucial for adapting to varying conditions. The clock's regulation is intrinsically linked to temperature responses, with C-Repeat/Dre Binding Factor (CBF) proteins playing a key role in cold adaptation. Environmental signals, such as light variations perceived by photoreceptors like Phytochrome B (PHY B) and Cryptochromes (CRYs), and temperature fluctuations, entrain the circadian oscillator, adjusting its phase and waveform [2]. Additionally, the overexpression of B-box (BBX) transcription factors, such as BBX18, BBX19, and BBX32, can alter the circadian cycle's duration, with implications for developmental processes like flowering [47].

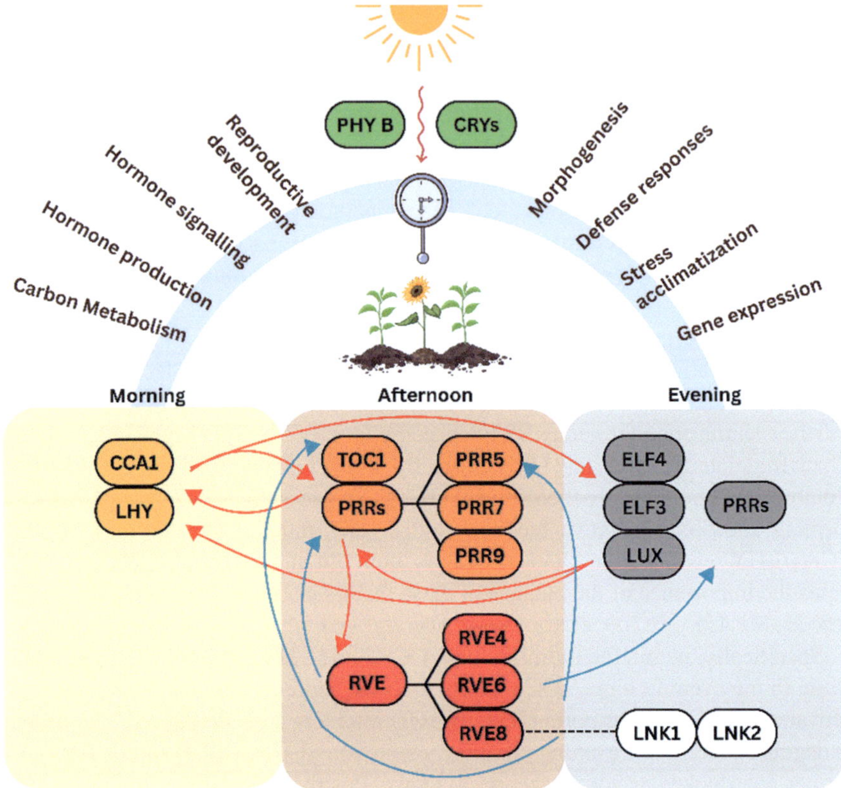

Fig. 1 Circadian rhythm in plants and key genes involved. The central loop features the morning-expressed MYB-like transcription factors Circadian Clock Associated I (CCA1) and Late Elongated Hypocotyl (LHY), which repress the afternoon-expressed Pseudo-Response Regulator (PRR) genes, including Timing of Cab Expression1 (TOC1), PRR5, PRR7, and PRR9. PRRs, and also regulate the evening complex genes Early Flowering3 (ELF3), ELF4, and Lux Arrhythmo (LUX), which are expressed at night and further repress morning and afternoon genes

3 Circadian Patterns in Rhizosphere Microbial Communities

Increasing research has demonstrated the presence of diurnal rhythms within rhizosphere microbial communities, highlighting a dynamic interplay between plants and their associated microorganisms. Initial studies in *Arabidopsis thaliana* and *Brachypodium distachyon* revealed significant variations in bacterial community composition between day and night, with approximately 3–14% of bacterial operational taxonomic units (OTUs) showing diurnal fluctuations in their relative abundances across various plant species, including rice and *Boechera stricta* [16, 28, 33]. Approximately 5% of bacterial OTUs and up to 12% of fungal OTUs displayed rhythmic fluctuations in relative abundance over a 24-h cycle. Importantly, a subset of these microbesmaintained rhythmicity even under constant light, a defining feature of endogenous circadian control. Specific bacterial taxa, including *Geobacter* and *Desulfosporosinus*, as well as fungal genera like *Penicillium* and *Pseudeurotium*, exhibited free-running rhythms independent of environmental cues, suggesting the presence of microbial-intrinsic clocks [5]. These findings have been further substantiated by observations of rhythmicity in both bacterial and fungal communities, with fungal communities, although showing a lower percentage of rhythmic taxa, exhibiting intense rhythmic changes in abundance [17].

Newman et al. made a significant contribution by establishing that these rhythms persist under constant conditions and are influenced by the plant's circadian clock. Disruptions in the *Arabidopsis* circadian oscillator, through overexpression or mutation of the LHY gene, resulted in alterations in the rhythmicity of the rhizomicrobiome community structure. This indicates that plant genotype influences the selection of niche-adapted microbial communities, with some OTUs displaying clock-dependent rhythmicity while others exhibit self-sustained endogenous timing [33]. The mechanisms behind circadian behaviors in rhizosphere microbes appear to be shaped by both intrinsic genetic regulation and environmental entrainment. While canonical circadian gene circuits have been characterized in model organisms, evidence is emerging that root-associated bacteria like *Pseudomonas fluorescens* and *Bacillus* spp. exhibit time-dependent gene expression profiles, particularly in genes linked to auxin biosynthesis, siderophore production, and quorum sensing—functions known to influence plant growth and nutrient uptake [31]. Environmental cues such as root exudate composition, which fluctuates over the diel cycle, act as potent entrainment signals. These exudates vary in quantity and chemical profile across day–night transitions, shaping microbial gene expression and community dynamics accordingly. Additionally, temperature and soil moisture—both rhythmic in natural systems—modulate microbial metabolic processes, including respiration and enzyme activity, potentially serving as secondary zeitgebers. While quorum sensing in rhizobacteria is well-known to coordinate collective behaviors like biofilm formation, recent studies suggest these signaling pathways may also participate in temporal synchronization, allowing microbial populations to collectively align their functional outputs with plant-derived rhythms [19]. Although the exact genetic components of microbial clocks in rhizosphere contexts are still being mapped, it is increasingly evident that both host-derived and environmental signals play essential roles in entraining microbial rhythms within the root-soil interface.

While most studies have focused on relative OTU abundance, recent work suggests that these rhythms reflect changes in microbial abundance and metabolic activity. Hubbard et al. demonstrated a threefold increase in total microbial DNA concentrations during the day compared to night, and Zhao et al. observed increased 16S rRNA abundance in daytime samples, indicating diurnal variations in transcriptional activity [17, 49].

The functional implications of these microbial rhythms are beginning to be understood. Metagenomic analyses and transcriptional profiling have revealed significant differences in carbohydrate, lipid, and amino acid metabolism between day and night-associated microbial communities. These metabolic shifts are correlated with changes in the abundance of specific bacterial groups, such as Burkholderiales, Caulobacterales, and Rhizobiales, as well as protists like Plasmodiophorida. Collectively, these studies underscore the intricate and dynamic relationship between plant circadian rhythms and the composition and activity of their rhizosphere microbial communities, highlighting the importance of temporal dynamics in plant–microbe interactions [4].

4 Interactions Between Plant and Microbial Circadian Rhythms

Rhizosphere-Specific Impacts of Plant's Circadian Clock

A plant's internal circadian clock significantly influences its association with its rhizomicrobiome. For instance, *Bacillus subtilis* can enhance nitrogen reductase activity in *Arabidopsis,* promoting plant growth. This interaction is facilitated by plants releasing organic compounds like amino acids, sugars, and organic acids through their roots, supporting microbial communities [24]. These organic compounds or a complex array of metabolites are biochemical signals—including flavonoids, terpenoids, and phytohormones such as auxin and salicylic acid, that not only shape microbial recruitment but also influence microbial gene expression in a time-sensitive manner [1, 12]. Shoaib et al. reported that changes in exudate profiles under drought stress modulated arbuscular mycorrhizal fungal activity and colonization rhythms, which in turn improved plant water uptake capacity and root system plasticity [40]. While initial studies identified only a handful of *Arabidopsis* phytochemicals exhibiting diurnal rhythms in soil accumulation, more recent research has revealed up to 50 such compounds in wild-type plants. Notably, mutants lacking functional TOC1 or CCA1 genes display reduced diversity in these root exudates, impacting the root microbiome. Soil conditioned by TOC1 and ZTL mutant plants resulted in decreased root microbiome diversity, and wild-type plants grown in this soil exhibited reduced biomass, demonstrating the direct link between circadian rhythm, exudate composition, and plant fitness [28].

The circadian clock's influence on root microbiome and exudate composition extends beyond general rhythms. The Casparian strip (CS), a root-specific barrier regulating water and solute export, including glucosinolates, is one such clock-controlled output [46]. Disruptions in the clock regulatory gene TIME FOR COFFEE (TIC) affect CS formation and alter rhizosphere bacterial populations. However, TOC1 knockdown did not affect CS integrity, suggesting that circadian effects on the root microbiome involve distinct pathways [10].

Further evidence of tissue-specific circadian control comes from the model legume *Medicago truncatula*. In this plant, root nodules, which house nitrogen-fixing bacteria, possess unique circadian clocks. While MtLHY, MtTOC1a, MtPRR5/9, and

MtLUX exhibit rhythmic expression in roots and shoots, only MtLHY shows rhythmicity in nodules. Disruption of MtLHY function led to fewer nodules and reduced shoot weight after inoculation with *Sinorhizobium meliloti*, highlighting the importance of nodule-specific circadian regulation [22]. Furthermore, nodulation itself influences the circadian rhythms of aerial tissues by altering nitrogen assimilation, causing a phase advance. This complex interplay between plant circadian rhythms, root exudates, and microbial interactions underscores the profound impact of the internal clock on plant health and interactions with its environment Fig. 2.

Feedback by Rhizosphere Microorganisms

The microbial communities in the rhizosphere exhibit distinct circadian rhythms, which significantly influence various physiological processes in plants. Khoshru et al. highlighted how auxin-producing rhizobacteria such as *Azospirillum* and *Bacillus* not only stimulate lateral root development but also entrain host physiological cycles through volatile signals and secondary metabolites, suggesting a feedback loop between microbial activity and plant developmental timing [20]. Temporal synchronization between plant and microbial metabolism extends into nutrient exchange processes. For instance, biological nitrogen fixation by rhizobia and phosphorus solubilization by phosphate-solubilizing bacteria are known to exhibit diel oscillations, often peaking during root exudation bursts that coincide with early morning or dusk periods. Barrios-Masias and Lazcano described how shifts in root carbon exudation across light–dark cycles modulate microbial nutrient-mineralization activity,

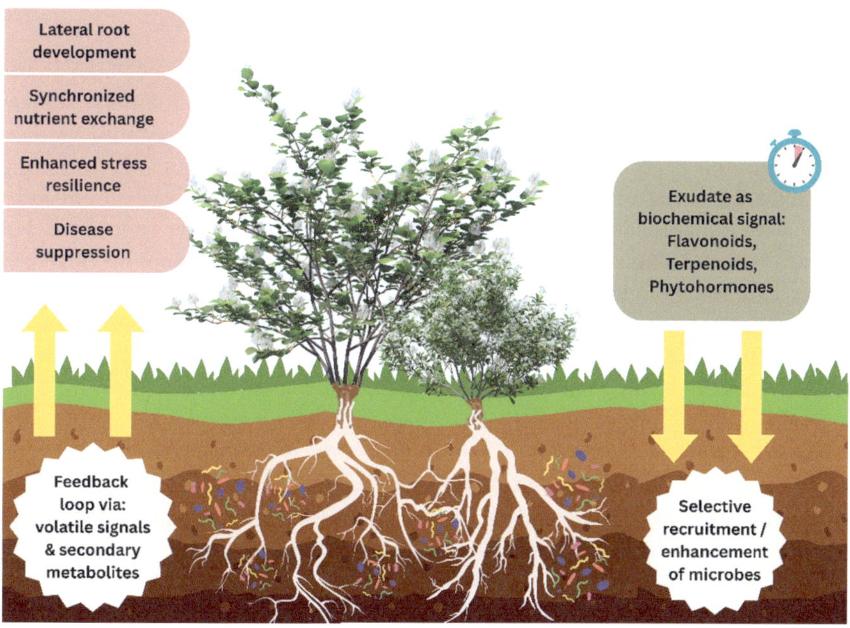

Fig. 2 Interplay of the circadian rhythm of the plants and their rhizomicrobiome

affecting N and P availability in vegetable crop systems [50]. Additionally, Mahawar et al. emphasized the role of synchronized microbial consortia, particularly plant growth-promoting rhizobacteria (PGPR) and arbuscular mycorrhizal fungi (AMF), in enhancing nutrient use efficiency, stress resilience, and disease suppression, especially when their activity aligns with the plant's endogenous rhythms [29]. Mycorrhizal fungi have been shown to peak in activity during specific periods when nutrient demand is high, facilitating the absorption of essential elements like nitrogen and phosphorus [44]. Such rhythmic nutrient solubilization not only bolsters plant health but also contributes to increased biomass production, particularly during critical growth phases.

Furthermore, specific bacterial populations in the rhizosphere can activate stress-response traits that help plants cope with adverse conditions. Studies have demonstrated that these microbes can produce osmoprotectants and phytohormones, such as abscisic acid (ABA), which are vital for regulating plant water stress responses [42]. The circadian rhythms of the rhizomicrobiome also influence plant immune responses. Timely microbial signaling can significantly enhance the expression of plant defense-related genes. For example, *Pseudomonas fluorescens* exhibits circadian patterns in its production of secondary metabolites that trigger plant immunity. This can thus effectively prime the plant's immune system, making it more capable of responding to pathogen attacks [34].

5 Applications and Implications

The circadian rhythm of the rhizomicrobiome holds transformative potential across agriculture, microbial engineering, ecosystem management, and climate adaptation. Understanding and leveraging these microbial oscillations as well as their synchrony with plants can optimize ecological and agronomic outcomes, offering an avenue for sustainable innovation. The synchronization of microbial circadian cycles with crop growth stages opens new dimensions in precision agriculture. For instance, *Pseudomonas fluorescens*, a common plant growth-promoting rhizobacterium, exhibits time-dependent expression of genes associated with auxin biosynthesis and siderophore production-critical for nutrient uptake and plant development [26]. Farmers could harness this knowledge by applying biofertilizers at dawn or dusk to match peak microbial activity. A practical example can be drawn from tea plantations, where time-aligned applications of microbial inoculants led to a 20–30% increase in nitrogen use efficiency [26]. Another study by [41] on indica rice cultivars demonstrated that root-associated microbial consortia showed rhythmic colonization behavior, which aligned with daily transpiration cycles, making them more effective during early morning hours [41].

The discovery that domestication has altered circadian clock genes in legumes—such as the ELF3 mutation in tropical soybean affecting flowering and potentially nodulation—suggests that clock regulation plays a critical role in symbiotic nitrogen fixation [37]. This indicates that disrupted circadian alignment between plant and rhizobium can impair nodule formation, highlighting a novel target for crop improvement. This offers a compelling proof of concept: breeding legumes with optimized circadian timing, or engineering rhizobia to synchronize with host rhythms, could significantly enhance nitrogen fixation across latitudes and reduce fertilizer dependency. In degraded soils, diurnal microbial behavior provides insight into the health of the soil ecosystem. Restoration strategies can include reintroducing native microbial consortia whose circadian timing supports stable nutrient cycling. [8] used the iRhizo-Chip to track *Bacillus subtilis* in real time, revealing rhythmic root colonization patterns that maintain soil pH and enzyme balance throughout the day [8]. Together, these insights underscore the vast potential of circadian-based strategies to fine-tune plant–microbe interactions across temporal, spatial, and environmental scales. With further research, integrating circadian biology into crop breeding, microbial engineering, and soil restoration has the potential to drive a paradigm shift in agricultural practices.

6 Challenges and Future Perspectives

Despite the promising advances in circadian microbiome research, several key challenges persist that hinder its complete translation into agricultural and ecological applications. One of the primary gaps lies in the limited understanding of circadian regulation among non-model microbial taxa, which make up most of the soil microbial diversity. Much of what is known about rhizomicrobial chronobiology originates from studies in well-characterized genera such as *Synechococcus* and *Bacillus*, leaving the vast majority of soil microbes unexplored regarding rhythmic gene regulation and metabolic cycling. Additionally, there is an incomplete integration of plant and microbial circadian rhythms; while plants have well-documented circadian systems, the temporal alignment—or misalignment—of microbial functions with plant rhythms remains poorly understood. From a technological perspective, real-time observation of microbial behavior in heterogeneous and opaque soil matrices continues to be a significant obstacle. Current methods such as metatranscriptomics and microfluidic sensors provide short-term insights under controlled conditions but often fall short in terms of scalability. The absence of long-term, high-resolution field studies has created a divide between controlled-environment findings and real-world agricultural applications, particularly in the context of fluctuating climates.

Looking ahead, a systems-level perspective is essential for modeling the feedback loops between circadian gene networks, root exudation rhythms, and microbial responses. There is also growing interest in exploring plant–microbe co-evolution through the lens of circadian biology, especially as preliminary evidence suggests that domestication and environmental adaptation have shaped temporal regulatory traits in both plants and their microbial partners.

7 Conclusion

In conclusion, the emerging field of rhizosphere microbiome circadian rhythms reveals a complex and dynamic interaction between plants and their microbial partners. Research consistently demonstrates that both bacterial and fungal communities in the rhizosphere exhibit diurnal fluctuations, influenced significantly by the host plant's internal clock. Perturbations in plant circadian rhythms directly impact the composition and rhythmicity of these microbial communities, highlighting the plant's role in shaping its surrounding microbiome. Beyond mere compositional changes, these rhythms are associated with variations in microbial abundance and metabolic activity, impacting essential functions like nutrient cycling and metabolism. The functional significance of these rhythms extends to plant fitness, as altered microbial dynamics can influence growth and stress responses. Ultimately, understanding the intricate interplay between plant circadian clocks and the rhizosphere microbiome offers valuable insights into optimizing plant health and productivity through targeted modulation of these rhythmic interactions.

References

1. Afridi MS, Kumar A, Javed MA, Dubey A, de Medeiros FHV, Santoyo G (2024) Harnessing root exudates for plant microbiome engineering and stress resistance in plants. Microbiol Res 279:127564. https://doi.org/10.1016/J.MICRES.2023.127564
2. Aguilar-Arnal L, Sassone-Corsi P (2015) Chromatin landscape and circadian dynamics: spatial and temporal organization of clock transcription. Proc Natl Acad Sci 112(22):6863–6870. https://doi.org/10.1073/PNAS.1411264111
3. Arana MV, Marín-De La Rosa N, Maloof JN, Blázqueza MA, Alabadí D (2011) Circadian oscillation of gibberellin signaling in Arabidopsis. Proc Natl Acad Sci USA 108(22):9292–9297. https://doi.org/10.1073/PNAS.1101050108/SUPPL_FILE/SD02.XLS
4. Baraniya D, Nannipieri P, Kublik S, Vestergaard G, Schloter M, Schöler A (2018) The impact of the diurnal cycle on the microbial transcriptome in the rhizosphere of barley. Microb Ecol 75(4):830–833. https://doi.org/10.1007/S00248-017-1101-0/FIGURES/2
5. Bending GD, Newman A, Picot E, Mushinski RM, Jones DL, Carré IA (2024) Diurnal Rhythmicity in the rhizosphere microbiome—mechanistic insights and significance for rhizosphere function. Plant, Cell Environ 48(3):2040. https://doi.org/10.1111/PCE.15283
6. Casal JJ, Qüesta JI (2018) Light and temperature cues: multitasking receptors and transcriptional integrators. New Phytol 217(3):1029–1034. https://doi.org/10.1111/NPH.14890
7. Creux N, Harmer S (2019) Circadian rhythms in plants. Cold Spring Harb Perspect Biol 11(9):a034611. https://doi.org/10.1101/CSHPERSPECT.A034611
8. Dai H, Wu B, Zhuang Y, Ren H, Chen Y, Zhang F, Chu C, Lv X, Xu J, Ma B (2024) Dynamic in situ detection in iRhizo-Chip reveals diurnal fluctuations of Bacillus subtilis in the rhizosphere. Proc Natl Acad Sci USA 121(40):e2408711121. https://doi.org/10.1073/PNAS.2408711121/SUPPL_FILE/PNAS.2408711121.SAPP.PDF
9. Duan Y, Wang T, Lei X, Cao Y, Liu L, Zou Z, Ma Y, Zhu X, Fang W (2024) Leguminous green manure intercropping changes the soil microbial community and increases soil nutrients and key quality components of tea leaves. Hortic Res 11(3). https://doi.org/10.1093/HR/UHAE018
10. Durr J, Reyt G, Spaepen S, Hilton S, Meehan C, Qi W, Kamiya T, Flis P, Dickinson HG, Feher A, Shivshankar U, Pavagadhi S, Swarup S, Salt D, Bending GD, Gutierrez-Marcos J (2021)

A Novel signaling pathway required for Arabidopsis endodermal root organization shapes the rhizosphere microbiome. Plant Cell Physiol 62(2):248–261. https://doi.org/10.1093/PCP/PCAA170

11. Farré EM, Weise SE (2012) The interactions between the circadian clock and primary metabolism. Curr Opin Plant Biol 15(3):293–300. https://doi.org/10.1016/J.PBI.2012.01.013
12. Gasperini D, Howe GA (2024) Phytohormones in a universe of regulatory metabolites: lessons from jasmonate. Plant Physiol 195(1):135. https://doi.org/10.1093/PLPHYS/KIAE045
13. Greenham K, McClung CR (2015) Integrating circadian dynamics with physiological processes in plants. Nat Rev Genet 16:10:598–610. https://doi.org/10.1038/nrg3976
14. Harmer SL (2009) The circadian system in higher plants. Annu Rev Plant Biol 60:357–377. https://doi.org/10.1146/ANNUREV.ARPLANT.043008.092054/CITE/REFWORKS
15. Hsu PY, Devisetty UK, Harmer SL (2013) Accurate timekeeping is controlled by a cycling activator in Arabidopsis. ELife 2013(2). https://doi.org/10.7554/ELIFE.00473
16. Hubbard CJ, Brock MT, Van Diepen LT, Maignien L, Ewers BE, Weinig C (2018) The plant circadian clock influences rhizosphere community structure and function. ISME J 12(2):400–410. https://doi.org/10.1038/ISMEJ.2017.172
17. Hubbard CJ, Harrison JG, McMinn R, Bennett Ponsford JC, Maignien L, Ewers B, Weinig C (2023) Rhizosphere microbial community composition shifts diurnally and in response to natural variation in host clock phenotype. MSystems 8(3). https://doi.org/10.1128/MSYSTEMS.01487-21/SUPPL_FILE/MSYSTEMS.01487-21-S0003.DOCX
18. Jamil F, Mukhtar H, Fouillaud M, Dufossé L (2022) Rhizosphere signaling: insights into Plant–Rhizomicrobiome interactions for sustainable agronomy. Microorganisms 10(5):899. https://doi.org/10.3390/MICROORGANISMS10050899
19. Kahl LJ, Eckartt KN, Morales DK, Price-Whelan A, Dietrich LEP (2022) Light/dark and temperature cycling modulate metabolic electron flow in pseudomonas aeruginosa biofilms. MBio 13(4). https://doi.org/10.1128/MBIO.01407-22/SUPPL_FILE/MBIO.01407-22-S0009.TXT
20. Khoshru B, Mitra D, Khoshmanzar E, Myo EM, Uniyal N, Mahakur B, Mohapatra PKD, Panneerselvam P, Boutaj H, Alizadeh M, Cely MVT, Senapati A, Rani A (2020) Current scenario and future prospects of plant growth-promoting rhizobacteria: an economic valuable resource for the agriculture revival under stressful conditions. J Plant Nutr 43(20):3062–3092. https://doi.org/10.1080/01904167.2020.1799004
21. Kim H, Kim Y, Yeom M, Lim J, Nam HG (2016) Age-associated circadian period changes in Arabidopsis leaves. J Exp Bot 67(9):2665–2673. https://doi.org/10.1093/JXB/ERW097
22. Kong Y, Han L, Liu X, Wang H, Wen L, Yu X, Xu X, Kong F, Fu C, Mysore KS, Wen J, Zhou C (2020) The nodulation and nyctinastic leaf movement is orchestrated by clock gene LHY in Medicago truncatula. J Integr Plant Biol 62(12):1880–1895. https://doi.org/10.1111/JIPB.12999/SUPPINFO
23. Kour S, Avatsingh AU, Sharma S, Singh N (2025) Applications and importance of metagenomic studies for exploring Rhizomicrobiome dynamics. Rhizomicrobiome Sustain Agric Environ, pp 373–389. https://doi.org/10.1016/B978-0-443-23691-4.00017-8
24. Lee S, Trịnh CS, Lee WJ, Jeong CY, Truong HA, Chung N, Kang CS, Lee H (2020) Bacillus subtilis strain L1 promotes nitrate reductase activity in Arabidopsis and elicits enhanced growth performance in Arabidopsis, lettuce, and wheat. J Plant Res 133(2):231–244. https://doi.org/10.1007/S10265-019-01160-4/FIGURES/7
25. Lei J, Jayaprakasha GK, Singh J, Uckoo R, Borrego EJ, Finlayson S, Kolomiets M, Patil BS, Braam J, Zhu-Salzman K (2019) Circadian clock-associated1 controls resistance to aphids by altering indole glucosinolate production. Plant Physiol 181(3):1344–1359. https://doi.org/10.1104/PP.19.00676
26. Liu M, Wang J, Li Z, Li X, Korpelainen H, Li C (2025) Circadian rhythms of microbial communities and their role in regulating nitrogen and phosphorus cycling in the rhizosphere of tea plants. Hortic Res 12(1). https://doi.org/10.1093/HR/UHAE267
27. Lu T, Ke M, Lavoie M, Jin Y, Fan X, Zhang Z, Fu Z, Sun L, Gillings M, Peñuelas J, Qian H, Zhu YG (2018) Rhizosphere microorganisms can influence the timing of plant flowering. Microbiome 6(1):231–231. https://doi.org/10.1186/S40168-018-0615-0

28. Lu T, Zhang Z, Li Y, Zhang Q, Cui H, Sun L, Peijnenburg WJGM, Peñuelas J, Zhu L, Zhu YG, Chen J, Qian H (2021) Does biological rhythm transmit from plants to rhizosphere microbes? Environ Microbiol 23(11):6895–6906. https://doi.org/10.1111/1462-2920.15820
29. Mahawar H, Bajpai A, Sreekanth D, Pawar D, Barman KK (2023) Emerging weeds under climate change and their microbial management. Bioinoculants: Biol Option Mitigating Glob Clim Chang, pp 57–86. https://doi.org/10.1007/978-981-99-2973-3_3
30. Marco S, Loredana M, Riccardo V, Raffaella B, Walter C, Nerva L (2022) Microbe-assisted crop improvement: a sustainable weapon to restore holobiont functionality and resilience. Hortic Res 9. https://doi.org/10.1093/HR/UHAC160
31. McClung CR (2019a) The plant circadian oscillator. Biology 8(1):14. https://doi.org/10.3390/BIOLOGY8010014
32. Millar AJ (2016) The intracellular dynamics of circadian clocks reach for the light of ecology and evolution. Annu Rev Plant Biol 67:595–618. https://doi.org/10.1146/ANNUREV-ARPLANT-043014-115619/CITE/REFWORKS
33. Newman A, Picot E, Davies S, Hilton S, Carré IA, Bending GD (2022) Circadian rhythms in the plant host influence rhythmicity of rhizosphere microbiota. BMC Biol 20(1):1–15. https://doi.org/10.1186/S12915-022-01430-Z/FIGURES/5
34. Nguyen NH, Trotel-Aziz P, Villaume S, Rabenoelina F, Schwarzenberg A, Nguema-Ona E, Clément C, Baillieul F, Aziz A (2020) Bacillus subtilis and Pseudomonas fluorescens Trigger Common and distinct systemic immune responses in Arabidopsis thaliana depending on the pathogen lifestyle. Vaccines 8(3):503. https://doi.org/10.3390/VACCINES8030503
35. Nohales MA (2021) Spatial organization and coordination of the plant circadian system. Genes 12(3):442. https://doi.org/10.3390/GENES12030442
36. Rawat R, Takahashi N, Hsu PY, Jones MA, Schwartz J, Salemi MR, Phinney BS, Harmer SL (2011) REVEILLE8 and PSEUDO-REPONSE REGULATOR5 form a negative feedback loop within the Arabidopsis circadian clock. PLoS Genet 7(3):e1001350. https://doi.org/10.1371/JOURNAL.PGEN.1001350
37. Rowson M, Jolly M, Dickson S, Gifford ML, Carré I (2024) Timely symbiosis: circadian control of legume-rhizobia symbiosis. Biochem Soc Trans 52(3):1419–1430. https://doi.org/10.1042/BST20231307
38. Saini R, Jaskolski M, Davis SJ (2019) Circadian oscillator proteins across the kingdoms of life: structural aspects. BMC Bio 17(1):1–39. https://doi.org/10.1186/S12915-018-0623-3
39. Sanchez A, Shin J, Davis SJ (2011) Abiotic stress and the plant circadian clock. Plant Signal Behav 6(2):223–231. https://doi.org/10.4161/PSB.6.2.14893
40. Shoaib M, Banerjee BP, Hayden M, Kant S (2022) Roots' drought adaptive traits in crop improvement. Plants 11(17):2256. https://doi.org/10.3390/PLANTS11172256
41. Singh A, Kumar M, Chakdar H, Pandiyan K, Kumar SC, Zeyad MT, Singh BN, Ravikiran KT, Mahto A, Srivastava AK, Saxena AK (2022) Influence of host genotype in establishing root associated microbiome of indica rice cultivars for plant growth promotion. Front Microbiol 13:1033158. https://doi.org/10.3389/FMICB.2022.1033158/XML/NLM
42. Singh A, Mazahar S, Chapadgaonkar SS, Giri P, Shourie A (2023) Phyto-microbiome to mitigate abiotic stress in crop plants. Front Microbiol 14:1210890. https://doi.org/10.3389/FMICB.2023.1210890
43. Venturi V, Keel C (2016) Signaling in the rhizosphere. Trends Plant Sci 21(3):187–198. https://doi.org/10.1016/J.TPLANTS.2016.01.005/ASSET/792ACBBC-68D7-4DA4-BC8F-2B6C4273D4E7/MAIN.ASSETS/GR1.JPG
44. Wahab A, Muhammad M, Munir A, Abdi G, Zaman W, Ayaz A, Khizar C, Reddy SPP (2023) Role of Arbuscular mycorrhizal fungi in regulating growth, enhancing productivity, and potentially influencing ecosystems under abiotic and biotic stresses. Plants 12(17):3102. https://doi.org/10.3390/PLANTS12173102
45. Wu JF, Tsai HL, Joanito I, Wu YC, Chang CW, Li YH, Wang Y, Hong JC, Chu JW, Hsu CP, Wu SH (2016) LWD-TCP complex activates the morning gene CCA1 in Arabidopsis. Nat Commun 7. https://doi.org/10.1038/NCOMMS13181

46. Xu D, Hanschen FS, Witzel K, Nintemann SJ, Nour-Eldin HH, Schreiner M, Halkier BA (2017) Rhizosecretion of stele-synthesized glucosinolates and their catabolites requires GTR-mediated import in Arabidopsis. J Exp Bot 68(12):3205–3214. https://doi.org/10.1093/JXB/ERW355
47. Yuan L, Yu Y, Liu M, Song Y, Li H, Sun J, Wang Q, Xie Q, Wang L, Xu X (2021) BBX19 fine-tunes the circadian rhythm by interacting with PSEUDO-RESPONSE REGULATOR proteins to facilitate their repressive effect on morning-phased clock genes. Plant Cell 33(8):2602–2617. https://doi.org/10.1093/PLCELL/KOAB133
48. Zhao X, Chen J, Guo M, Li C, Hou N, Bai S (2022) Constructed wetlands treating synthetic wastewater in response to day-night alterations: performance and mechanisms. Chem Eng J 446:137460. https://doi.org/10.1016/J.CEJ.2022.137460
49. Zhao K, Ma B, Xu Y, Stirling E, Xu J (2021) Light exposure mediates circadian rhythms of rhizosphere microbial communities. ISME J 15(9):2655–2664. https://doi.org/10.1038/S41396-021-00957-3
50. Zhao M, Zhao J, Yuan J, Hale L, Wen T, Huang Q, Vivanco JM, Zhou J, Kowalchuk GA, Shen Q (2021) Root exudates drive soil-microbe-nutrient feedbacks in response to plant growth. Plant, Cell Environ 44(2):613–628. https://doi.org/10.1111/PCE.13928

Chapter 5
Circadian Rhythms in the Algal World

Atreyee Bagchi, Deepanshu Vasuja, and Aditi Kothari Chhajer

1 Introduction

Far away on a remote coastline, hidden between mountains reaching for the twinkling stars, a team of marine biologist's stumble across an unexpected sight. Under the white glow of the full moon, vast sheets of algae shone as one large mythical creature, as if upon the order of an invisible conductor. Night after night, the glow followed a strict schedule—never early, never late. The scientists, now intrigued by the fascinating time adherence, began tracking the algae's biological processes such as growth, reproduction, and even their resting state. These microscopic organisms, seemingly simple, are connected to the grand clock of nature—linked to the seasons, the moon, the tides, and even the unseen ticking of internal rhythms. These are not mere ocean drifters but creatures who view punctuality as their only way of life for survival.

But what controlled these rhythms? And how did algae—one of Earth's most primitive inhabitants—master the art of timekeeping long before humans even understood the concept of calendars? The astonishing realm of algal periodicity, where biological and natural factors combine to create a captivating dance of cycles and signals, is far deeper than just glowing in the dark. It is an ode to functioning and a habit of life.

1.1 What Are Rhythms?

All organisms, from the smallest bacteria to the largest mammals, operate an intrinsic rhythmic process that regulate various biological activities. Just as humans tend to

A. Bagchi · D. Vasuja · A. K. Chhajer (✉)
Sri Venkateswara College, Delhi University-South Campus, Delhi, India
e-mail: aditi@svc.ac.in

A. Kothari Chhajer and V. Singh Gour (eds.), *Introduction to Rhythms in Plants*, Studies in Rhythm Engineering, https://doi.org/10.1007/978-981-95-3592-7_5

sleep at night and wake with the sunrise, these rhythms ensure that specific functions occur at designated times throughout the day. A biological rhythm is defined as a process that repeats at consistent intervals, systematically fluctuating between peaks (maximum values) and troughs (minimum values). For example, the release of the hormone melatonin, which induces sleep in humans, rises to its highest levels at night and declines during the day, which is why we tend to feel sleepy as night approaches. The difference between these highs and lows, known as amplitude, reflects the extent of variation in the process, while the time between consecutive peaks (or troughs) defines the cycle's period (T).

A key aspect of these rhythms is the phase shift—a modification in the timing of the cycle that causes critical points like peaks, troughs, or transitions to occur earlier or later than expected. Such shifts are essential for enabling organisms to adjust to environmental cues such as light–dark cycles or feeding times. For instance, a phase advance, often triggered by exposure to light earlier than usual, can cause an organism to wake up earlier than usual. In contrast, a phase delay, which might result from late-night light exposure, pushes events like sleep onset to a later time. Such changes are observed when we experience jet lags while traveling across countries. Understanding these adjustments is central to circadian biology, as they illustrate how external factors like light and temperature help synchronize an organism's internal clock with its environment.

1.2 Properties of Rhythms

To appreciate the importance of these rhythms, it is important to understand its fundamental properties.

1. Innate behavior—Circadian rhythms are intrinsic and self-sustaining. This means that an organism's internal clock continues to operate throughout its life, even in the absence of external cues such as light or darkness. Thus the rhythms is also endogenously controlled in the organism rather than solely dependent on external conditions [1].
2. Entrainment—While the core circadian rhythm is generated internally, it is highly adaptable. Entrainment is the process by which zeitgebers, such as the light–dark cycle—adjust and synchronize the internal clock to the environment. This fine-tuning can be understood through phase advance and phase delay. A phase advance occurs when exposure to light early in the day shifts the internal clock of algae to an earlier time, allowing them to begin their physiological activities sooner. Conversely, a phase delay happens when exposure to light in the evening pushes the clock to a later time, delaying the onset of sleep. These adjustments ensure that the rhythm remains aligned with the natural cycle of the environment, which is crucial for survival.

3. Temperature Compensation—In many biological processes, increasing temperature speeds up chemical reactions. However, circadian rhythms display temperature compensation, meaning that their period (or cycle length) remains relatively constant despite fluctuations in ambient temperature. This feature is critical because it allows the internal clock to maintain a consistent timing mechanism regardless of environmental temperature changes, ensuring stable regulation of physiological processes [1, 2].
4. Free-Running Period—The free-running period is the natural cycle length of a circadian rhythm in the absence of external time cues. In other words, if an organism is kept in a constant environment (for instance, continuous dim light), its internal clock will run at its own pace [1].

Other biological rhythms such as circalunar (regulated by the phases of moon, eg. phenomenon of tides), ultradian (rhythm with time period less than 24 h, eg. chlorophyll a level in *C. reinhardtii*), infradian (rhythm with time period more than 24 h but shorter than circannual cycle, e.g., vegetative growth of *Ulva lactuca*) [3], and diurnal rhythms have also evolved in organisms to regulate specific processes to support life.

While both circadian and diurnal rhythms relate to a roughly 24-h cycle, the key difference lies in the fact that circadian rhythms are generated by an internal clock and persist under constant conditions (they are endogenous). Even when external cues are removed, these rhythms continue to operate with some deviations. Diurnal rhythms, on the other hand, refer to patterns of activity that occur during the day, which are typically driven by external factors like daylight. Although diurnal behaviors often reflect the influence of the circadian system, they are not solely produced by it; they can be a direct response to environmental cues. For example, *Acetabularia* under constant environmental conditions—say, continuous dim light—exhibits a self-sustained, intrinsic circadian rhythm that governs its photosynthetic activity. This rhythm, characterized by a free-running period (its natural cycle length when isolated from external cues), might deviate slightly from a 24-h cycle. For instance, in the absence of a natural light–dark cycle, the peak of photosynthetic activity may shift gradually each day. However, in its natural environment, the regular cycle of daylight and darkness acts as a powerful zeitgeber, entraining the alga's internal clock to synchronize with the 24-h day. This entrainment ensures that photosynthesis occurs predominantly during the day—creating a diurnal pattern—while other processes such as cellular repair or nutrient uptake may be optimized during the night. Additionally, the mechanism is temperature compensated, meaning that fluctuations in ambient temperature do not significantly alter the rhythm's period. Thus, while the core circadian rhythm in *Acetabularia* is inherently generated and maintained, environmental cues refine it to produce predictable diurnal cycles that align with the Earth's natural cycles.

It is understood from above that all these rhythms are a function of external factors which then, at genetic level, regulate processes. In other words, these rhythms are biologically regulated by an endogenous clock or oscillator—an internal mechanism present in all organisms that governs the timing and intensity of various biological processes. The concept of a circadian system consisting of three principal elements: input, central oscillator, and output, was first proposed by [4, 5], which explains that the oscillator interprets external cues like sunlight and translates them into metabolic responses by modulating biomolecular levels [6].

While the circadian clock functions as a self-sustaining oscillator with a cycle close to 24 h, its period is not exactly one day. As a result, regular adjustments, or resetting, are crucial for the clock to synchronize with daily environmental changes. Light serves as one of the primary cues for this resetting process. For instance, in the phase-resetting mechanism *of C. reinhardtii,* light exposure—especially blue, green, and red wavelengths—resets the circadian clock by triggering the rapid degradation of the clock protein ROC15 via a proteasome-dependent mechanism (a cellular process that removes unwanted or damaged proteins using a specialized protein complex called the proteasome. This system maintains protein balance by breaking down misfolded or excess proteins. This process requires the F-box protein ROC114, which facilitates ROC15 degradation. Mutations in ROC114 hinder this breakdown, and its reintroduction in mutants restores normal degradation. Moreover, cells lacking ROC15 display defective light-induced phase resetting, with red light proving to be particularly effective in driving the response [7].

Another important component of the circadian rhythm is the feedback loop. Complex feedback loops are a key part of the daily timing system. In animals, several loops have been identified that send signals from the outputs of the circadian system back to the core clock, and similar loops likely exist even in single-celled organisms [8]. These feedback loops keep the circadian rhythms in check by constant regulation.

1.3 Algal Rhythms

Every organism operates on its own internal timetable, coordinating daily routines that not only keep life running smoothly but also enhance survival—almost like a natural, built-in discipline. Thus, the earlier mentioned *dance of cycles and signals* is not a recent discovery. In fact, such rhythmic synchronization evolved nearly two billion years ago in unicellular marine organisms, allowing them to align their biological functions with the Earth's rotation—establishing the 24-h daily cycle—and its revolution around the sun, shaping the approximately 365-day annual cycle. These astronomical and geophysical motions give rise to two primary types of biological rhythms: the daily circadian rhythms and the yearly circannual rhythms, respectively [9]. These rhythms are crucial because they regulate nearly every function in an organism—from simple activities like waking up in the morning to complex processes like digestion.

In algae, these age-old rhythms orchestrate not only their metabolism but also their reproduction and other functions. One striking example is the circadian rhythm of buoyancy observed in the green macroalga *Aegagropila linnaei* when it forms spherical aggregates—dense, ball-like clusters of filaments that enhance survival in aquatic environments. This daily cycle allows the algae to float toward the water's surface during daylight hours, optimizing light exposure for photosynthesis, and sink at night, potentially aiding in nutrient absorption or protection from predators [10].

Similarly, the movement patterns of *C. reinhardtii*, a unicellular green alga, are governed by its internal circadian clock. This regulation ensures that its tactic movements—directed responses to environmental stimuli such as light (phototaxis) and gravity (gravitaxis)—are synchronized with the day-night cycle. By adjusting its movement accordingly, *Chlamydomonas* can maximize energy capture during daylight while potentially avoiding harmful light intensities or unfavorable conditions. These examples highlight the profound role of biological rhythms in coordinating essential activities across diverse algal species.

Rhythms of cell division, UV light sensitivity and starch accumulation have also been observed [10]. These rhythms, in algae, not only synchronizes light-dependent processes but also finely tunes cellular metabolism to optimize resource use. For example, nitrate reductase—the key enzyme responsible for converting nitrate (NO_3) into nitrite during nitrogen assimilation—is tightly regulated by the circadian clock. By aligning the enzyme's activity with the daily light–dark cycle, algae ensure that nitrate conversion occurs at the most beneficial time, thus enhancing metabolic efficiency. Moreover, the availability of nitrate itself can influence the rhythm, suggesting a feedback loop in which the substrate acts as a signal to adjust the circadian system. This dynamic interplay helps algae coordinate nutrient uptake with environmental conditions, maximizing growth and energy utilization [8].

1.4 Understanding the Circa World

Behind our daily routines, a hidden clock synchronizes our sleep, energy levels, and hormone release with the natural cycle of light and darkness. This internal timekeeper—known as the circadian rhythm (derived from the Latin *circa diem*, "about a day"), a term coined by Franz Halberg in 1959—refers to a cycle that repeats approximately every 24 h ($T = 24$ h) and governs our daily activities in tune with the day/night cycle. While these rhythms are inherently generated by our biology, under natural conditions, they are profoundly influenced by external cues, or zeitgebers, such as photoperiod, temperature, and nutrient availability, with light being the most important zeitgeber. For instance, *Acetabularia*, a green alga historically significant in circadian biology, exhibits strong and persistent rhythms of photosynthesis even in enucleated cells. Under normal conditions, light triggers gene transcription in the nucleus, leading to the production of mRNAs that encode proteins essential for photosynthesis. Remarkably, when the nucleus is removed or RNA synthesis is inhibited, photosynthesis continues unabated suggesting that pre-existing, stable cytoplasmic mRNAs maintain the rhythmic cycle without the need for new transcription [11].

Another compelling example of environmental cues influencing biological processes is the seasonal green tide dominated by *Ulva prolifera*. Due to its rapid growth and reproductive capabilities, this alga appears and disappears between spring and autumn. However, its balance between vegetative growth and reproduction is highly sensitive to factors such as high light intensity, elevated temperatures, grazing, physical damage, and eutrophication. These stressors can shift biomass allocation from growth toward reproduction, triggering sporogenesis and forming a spore bank that sets the stage for future green tide events. Thus, understanding how these environmental factors influence spore formation in *U. prolifera* is essential for predicting and managing these blooms [12].

While daily environmental cues influence immediate physiological responses, longer-term seasonal patterns are governed by a different kind of internal rhythm known as the circannual rhythm. Unlike circadian rhythms, which operate on a roughly 24-h cycle, circannual rhythms regulate the timing and duration of an organism's major life phases over the course of a year. This annual clock determines when an organism enters its reproductive phase, when it should allocate energy toward vegetative growth, and when to enter dormancy, ensuring that these transitions occur during optimal environmental conditions. In this way, the circannual rhythm provides a broader temporal framework that complements the finer daily adjustments made in response to zeitgebers.

Furthermore, research on the temporal organization of circannual timing has revealed three fundamental patterns.

Trans-Generational: This pattern occurs in short-lived species—such as certain marine algae and insects that produce multiple generations within a year (e.g., some migratory butterflies and moths). In these organisms, the intrinsic circannual timing mechanism is transmitted from one generation to the next and is typically expressed during a specific phase of the life cycle (for example, the excystment phase in algae). Since each individual cyst or unit undergoes this process only once, the response is observed at the population or cohort level.

Modular pattern which is found in longer-lived insects and is characterized by a progression through discrete developmental stages—from egg to larva, then pupation, and finally the emergence of a sexually mature adult.

Progressive pattern which involves continuous development from juvenile stages through adulthood and into old age [1].

Together, these patterns illustrate the diverse strategies by which organisms synchronize their life cycles with environmental cycles, ensuring optimal timing for growth, reproduction, and survival. The marine algae *Alexandrium tamarense* (earlier known as *Gonyaulaux*) is one of the best unicellular models to understand circannual rhythms [1]. Several advantages associated with the model include a prominent alteration between growth and replication during summers, when the algae reside at sea surface and produce algal bloom, and prolonged wintering dominant phase. In autumn, the haploid cells convert to diploid planozygote which mature into dense resistance cyst that sink into the depths of the ocean [1] (Fig. 1).

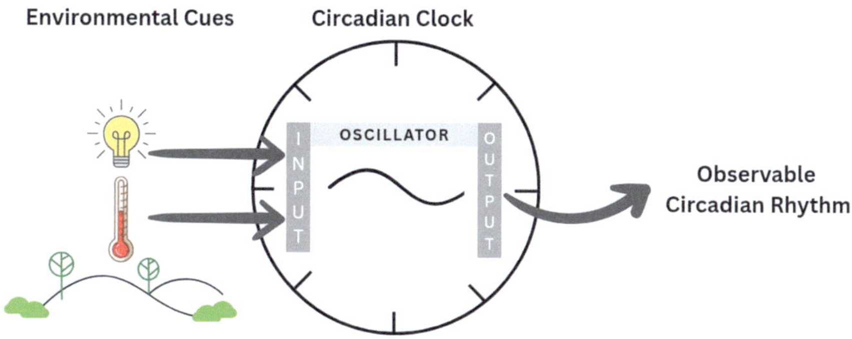

Fig. 1 Diagrammatic representation depicting environmental cues and components of circadian clock-input, oscillator, and output—which result in observable circadian rhythms. Rhythms are biologically regulated by an endogenous clock or oscillator—an internal mechanism present in all organisms that governs the timing and intensity of various biological processes. The external factors or zeitgebers act as the input which are processed by the said oscillator. The result or output is the various observable circadian rhythms seen in the organism

2 Evolution of Rhythms in Algae

Cyanobacteria were the first prokaryotes found to possess a circadian clock, a feature that is crucial for their survival. They are exceptionally adaptable prokaryotes that have evolved efficient mechanisms for environmental adaptation [5]. Throughout the course of evolution, certain components of the circadian rhythm have remained conserved along the green lineages including in the prokaryotes. The difference arises due to increasing complexity of organisms requiring the involvement of more genes for proper functioning. For instance, clock complexity increases from picoalga like *O. tauri* to flagellate algae *C. reinhardtii* and from mosses to angiosperms [10]. The fundamental elements of the circadian clock, such as PRR (Pseudoresponse Regulators) and RVE (REVEILLE) family of genes, are conserved throughout the green lineage. The Evening Complex (EC) is a key regulator of the circadian clock in angiosperms (flowering plants), consisting of proteins like ELF3 (EARLY FLOWERING 3) and ELF4 (EARLY FLOWERING 4). The fact that ELF3 and ELF4 are largely conserved in streptophytes (a group that includes land plants and their closest algal relatives) suggests that these genes have been maintained throughout evolution. This conservation indicates their fundamental role in circadian regulation, even before plants fully adapted to terrestrial environments [10].

However, some changes in the presence or absence of specific genes, especially between algae and land plants are worth noting. For example, ZTL (ZEITLUPE) and GI (GIGANTEA) are found in most streptophytes but are absent in select chlorophyte algae; CRY (Cryptochrome) variety and function differ between algae & land plants. LUX (LUX ARRHYTHMO) expression patterns differ between *A. thaliana* & *C. reinhardtii* [10]. The output of circadian rhythm varies in different organisms, in land plants, the rhythm is observed in leaves, flower and stomatal movements whereas in unicellular algae, it is observed in movement and nutrient intake.

While these internal clocks regulate vital processes in many organisms, even simpler life forms like algae have evolved unique strategies to harness environmental cues. Algae, unlike humans, lack specialized organs like eyes to detect light. Instead, they rely on a process called photoperiod sensing, in which photoreceptors monitor light and work in tandem with the circadian clock to keep track of time. This mechanism has grown so essential that the circadian clock governs 90% of the transcriptome [13]. Another interesting point to note is that plants and algae have a gene CONSTITUTIVE PHOTOMORPHOGENIC 1 (COP1) which is active in dark and inhibits photo-periodically controlled pathways. During the day, COP1 is inhibited by photoreceptors allowing activation of these pathways [13].

In *Euglena gracilis*, physiological studies indicate that cell cycle progression is regulated by two distinct circadian rhythms. The first mechanism ensures that in photoautotrophic cultures, the decision to progress through the cell cycle depends on circadian phase-dependent sensitivity to photo induction. While cell cycle progression is halted in the dark, G2-phase cells can only proceed to mitosis if they receive light exposure at subjective dusk. The second mechanism regulates population growth by preventing G2 cells from entering mitosis during daylight hours. Cyclic adenosine monophosphate (cAMP) is believed to play a role in the circadian control of the *E. gracilis* cell cycle [11].

Another study focused on the translational control of the bioluminescent system in *Gonyaulax polyhedra* by examining the Luciferin Binding Protein (LBP). Researchers found that an RNA-binding protein called the Circadian Controlled Translational Regulators (CCTRs) binding activity diminishes at night—when LBP synthesis is initiated—and increases during the day when LBP production falls. This pattern of regulation helps explain why bioluminescence is predominantly observed at night [14].

A fascinating fact about circadian rhythms is that they influence more than just major processes like reproduction—they also regulate activities inside cells. For instance, within the nucleus, essential functions such as transcription and DNA repair follow a circadian pattern. The nucleus also contains various DNA repair systems. One key system, nucleotide excision repair (NER), which fixes single-strand DNA damage and UV-induced lesions, varies in both activity and efficiency throughout the day. Overall, the circadian clock plays a significant role in controlling transcription, managing DNA damage, and organizing nuclear architecture [15].

Similarly, mitochondria, which are central to energy production and redox balance, are closely linked with the circadian clock. Their ability to perform oxidative functions follows a daily rhythm. This rhythmic behavior likely comes from the direct control of many nuclear genes that code for mitochondrial proteins by the core clock components. As a result, the expression of these genes aligns with the daily fluctuations seen in the corresponding mitochondrial proteins [15].

3 Understanding the Role of Rhythms Through Various Algal Models

Algal models are invaluable for circadian research, as they reveal how biological clocks link rhythmic gene expression to cellular functions. Studies in systems from cyanobacteria and microalgae to *Acetabularia*, as mentioned in the beginning, have illuminated key features like temperature compensation, phase shifts, and the interplay between transcription and metabolism, deepening our understanding of growth, metabolism, and adaptation. Let's highlight some more algal models.

3.1 C. reinhardtii—*A Model Algal System*

C. reinhardtii is a single-celled alga characterized by its large U-shaped chloroplast and two flagella. It also possesses a basic visual system known as the eyespot apparatus, which is found at the periphery of the chloroplast. This alga presents experimental benefits akin to those of yeast, but unlike yeast, it can be utilized for various processes central to this green alga [16] due to which *C. reinhardtii* is sometimes referred to as "green yeast" [7]. It is classified as a haplont (an organism that spends most of its life cycle in the haploid (n) stage, with a brief diploid (2n) phase limited to the zygote before meiosis restores haploidy), and the creation of both plus and minus gametes can be triggered by nitrogen depletion in light.

Circadian rhythms play multiple roles in algae, particularly in *C. reinhardtii*. For instance, its cell division is regulated by the circadian clock, occurring at the onset of subjective night [11]. Additionally, *C. reinhardtii* exhibits circadian-dependent variations in UV radiation sensitivity, which negatively impacts nuclear DNA replication. Its cells exhibit a survival rhythm in response to UV radiation, with peak sensitivity occurring at the end of the day and the beginning of the night, coinciding with nuclear division. This rhythm persists under constant light, confirming its regulation by the circadian clock [17]. In the 1970s, Victor Bruce showed that these algae swam toward a light source, achieving the highest accumulation during daylight hours.

UV radiation poses a significant threat to *Chlamydomonas* cells, as they are frequently exposed to UV-B in both soil and shallow aquatic environments. Even in aquatic habitats, UV can penetrate deeply [17]. The potential impact of UV irradiation likely influenced the evolution of circadian clocks, particularly given the absence of the ozone layer during early cellular evolution. This supports the theory that the daily UV cycle may have driven the development of circadian systems, further reinforced by the discovery of cryptochromes—proteins homologous to DNA photolyase—that aid in UV-induced DNA damage repair [17].

Beyond UV protection, circadian rhythms regulate key physiological behaviors in *C. reinhardtii*. For example, circadian rhythms influence nutrient acquisition strategies. At night, *C. reinhardtii* exhibits increased chemotaxis toward ammonium, despite ammonium uptake occurring only at dawn. Similarly, nitrite uptake and nitrite reductase activity peak at dawn [18]. These activities are temporally distinct: phototaxis is active during the day to enhance photosynthesis, while nighttime chemotaxis ensures nutrient acquisition in the absence of sunlight.

3.1.1 Why is *C. reinhardtii* Considered a Model Organism?

The unicellular chlorophyte algae, *C. reinhardtii* has been a popular model for circadian biology, even for genetic studies, due to its various measurable circadian-controlled physiological responses. It has a straightforward life cycle, facilitates easy mutant isolation, and provides an expanding set of tools and techniques for molecular genetic research [19]. Another key feature of this algae is its ability to grow using light as its sole energy source (photoautotrophic growth) or on acetate in the dark (heterotrophic growth), enabling detailed studies of genes and proteins essential for photosynthesis and respiration [20]. Although yeast also serves as a good model system to study rhythms, but due to the presence of chloroplast and flagella in *Chlamydomonas*, it serves as a better model for studying circadian rhythms in simple eukaryotes.

3.1.2 Clock Components in Algae

While the fundamental features of circadian clocks are similar across various model organisms, the actual sequences of the clock genes and proteins differ significantly between different biological kingdoms. A critical component of the circadian clock's central oscillatory system is a feedback loop mechanism that regulates the expression and translation of certain genes, known as "clock genes." [21]

Clock mutants were initially identified in *Drosophila melanogaster* and *Neurospora crassa*, and shortly thereafter, in the green alga *C. reinhardtii*. The circadian clock components in *C. reinhardtii* were found to [22] include various key proteins that regulate the algae's rhythmic behavior, especially in relation to chloroplast bioluminescence. These components include the RNA-binding protein complex **CHLAMY1**, a casein kinase and several proteins referred to as RHYTHM OF CHLOROPLAST **(ROC)** proteins [23, 24].

In *C. reinhardtii*, circadian clock components regulate various cellular processes, with one of the most critical functions being the maintenance of chloroplast rhythmicity. This rhythmicity is regulated by a range of biological processes, including flagellar activity, the ubiquitin–proteasome system, transcription and transcript metabolism, gene silencing, membrane transport and trafficking, signal transduction, responses to DNA damage, and apoptosis [22] (Table 1).

Genome sequence annotation and comparative analyses of conserved domains have enabled the categorization of candidate clock-associated genes into distinct biological processes. These findings underscore the complexity of chloroplast rhythmicity, and the multiple cellular pathways involved in sustaining circadian regulation. Further analyses of *C. reinhardtii*'s clock-associated genes revealed that all candidate

Table 1 Genes associated with a range of biological processes in *C. reinhardtii*

Biological process exhibiting rhythmicity	Associated Genes in *C. reinhardtii*
Flagellar function	FAP256, ODA4, PF9, DLC7a, FLA14
Ubiquitin–proteasome system	SKP1, ROC114
Transcription and transcript metabolism	ROC15, ROC40, ROC56, ROC59, ROC66, ROC75, ROC76, ROC93, XRN1
Membrane trafficking and transport	VPS11, ROC81, DRP2
Signal transduction	MAPKKKK
DNA damage response	ATR1
Apoptosis	ALIX

FAP—Flagellar Associated Protein; **ODA4**—Organelle Genome Database for Algae 4; **PF**—Paralyzed Flagella; **DLC**—Dynein Light Chain; **FLA—**Fasciclin-like arabinogalactan protein; **SKP**—S-phase kinase-associated protein; **ROC**—RHYTHM OF CHLOROPLAST; **XRN**—5′–3′ Exoribonuclease; **VPS**—Vacuolar Protein Sorting-associated Protein; **DRP**—Dystrophin-related protein; **MFT**-; **MAPKKKK**—Mitogen-Activated Protein Kinase Kinase Kinase Kinase; **ATR**—Ataxia telangiectasia and Rad3-related protein; **ALIX**—ALG-2-interacting protein X.

genes, except ROC40, contain introns that follow the GT–AG splice rule. Additionally, alternative splicing events have been observed upstream of the predicted initiation codon in the ROC75 gene. These variations suggest that post-transcriptional regulation plays a role in modulating circadian functions in *C. reinhardtii* [22] (Fig. 2).

Homology with Other Species

Circadian rhythms in *C. reinhardtii* are regulated by a network of conserved proteins, many of which share homology with clock components in other species. Understanding these homologous relationships is crucial for deciphering the evolution and function of circadian systems in green algae. For instance, a group of highly conserved proteins, including **Ser-/Thr kinases** and **protein phosphatases (PPs)**, has been maintained in *C. reinhardtii.* Among these highly conserved proteins, **CASEIN KINASE 1** and **2 (CK1 and CK2)**, along with a protein kinase known to phosphorylate key clock components in *Neurospora, Drosophila*, and mammals—have been detected. Additionally, the protein phosphatases **PP1** and **PP2A**, which counteract kinase activity, are well conserved in *C. reinhardtii* [16].

A comparative genomic analysis between *C. reinhardtii* and *Volvox carteri* has revealed six conserved gene loci of the **ROC** protein family: **ROC15, ROC40, ROC55, ROC66, ROC75**, and **ROC114** [22].

ROC40 features a single **MYB** repeat, similar to clock proteins such as **CIRCADIAN CLOCK ASSOCIATED 1 (CCA1)** and **LATE ELONGATED HYPOCOTYL (LHY)**. It also shares structural resemblance to CCA1 and LHY beyond the MYB domain. ROC66, in contrast, contains a CCT domain akin to CONSTANS (CO), CONSTANS-like (COL), and TIMING OF CAB EXPRESSION 1 (TOC1) proteins. Additionally, ROC66 possesses a single repeat of the B-box zinc-finger domain, commonly found in CO and COL proteins, along with an incomplete

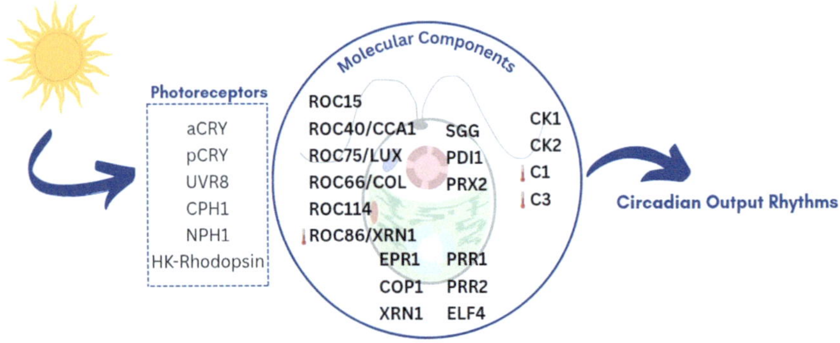

Fig. 2 Molecular components of the circadian clock of flagellate alga *C. reinhardtii*. Clock elements capable of incorporating temperature data and potentially playing a role in temperature entrainment are marked with a thermometer. pCRY is known to have dual roles both in input pathway and also in oscillator [58]. The clock components include: aCRY—animal-like CRYPTOCHROME; pCRY—plant CRYPTOCHROME; UVR-8—UV RESISTANCE LOCUS 8; CPH1—Cryptochrome Photolyase Homolog 1; NPH1—Phototropin [11]; HK-Rhodopsin—Histidine Kinase Rhodopsins; ROC—RHYTHM OF CHLOROPLAST; CCA1—CIRCADIAN CLOCK-ASSOCIATED1; LUX—LUX ARRHYTHMO; COL—CONSTANS-like; XRN1—5'-3' EXORIBONUCLEASE 1; SGG—Starch Synthase Glucan; PDI1—Protein Disulfide Isomerase 1; PRX2—2-Cys Peroxiredoxin; CK—CASEIN KINASE; C1—subunit 1 of the RNA-binding protein CHLAMY1; C3—subunit 3 of the RNA-binding protein CHLAMY1; EPR—Expressed Protein in Response, also known as Phosphate Starvation Response; COP—CONSTITUTIVE PHOTOMORPHOGENIC; PRR—PSEUDO RESPONSE REGULATOR; ELF—EARLY FLOWERING

second B-box, a feature observed in certain COL proteins **(COL9, COL10, COL11, and COL12)** [22, 24].

Specifically, ROC15 and ROC75 encode GOLDEN2/ARR/Psr1 (GARP) transcription factors, resembling LUX (PCL1) and BROTHER OF LUX ARRHYTHMO aka BOA (NOX) proteins in *Arabidopsis thaliana*, which interact with ELF3 and ELF4 and form a complex to regulate the circadian clock by promoting CCA1 expression [25]. Likewise, ROC40 encodes a MYB transcription factor homologous to CCA1/LHY, which are crucial regulators of the morning phase in the *A. thaliana* circadian clock. Furthermore, ROC66 encodes a CONSTANS-like (COL) protein that significantly influences the *C. reinhardtii* clock [22]. However, in *A. thaliana*, misexpression of COL family genes results in only a limited effect on the circadian clock [26].

While *C. reinhardtii* shares homologous genes with *A. thaliana*, its circadian clock functions under distinct regulatory mechanisms, making direct application of the *A. thaliana* model to *C. reinhardtii* challenging [27, 28]. The DNA-binding motifs of the ROC proteins *in C. reinhardtii* share similarities with *A. thaliana* circadian clock and photoperiodic flowering regulators, further supporting a conserved clock regulation mechanism across species.

In addition to these conserved genes, several genes in *C. reinhardtii* contribute to its circadian clock but lack homologs in *A. thaliana*. These include ROC55, ROC59,

ROC114, NAT3, XRN1 (ROC86), C1, C3, and pCRY, underscoring the unique regulatory network governing *C. reinhardtii's* circadian rhythms [22, 29–32].

3.1.3 Understanding/Decoding the ROC Family

The ROC family of proteins includes putative DNA-binding proteins such as ROC15, ROC40, ROC66, and ROC75, as well as an F-box protein (ROC114) and a leucine-rich repeat protein (ROC55). As mentioned previously, ROC15 and ROC75 encode GARP transcription factors similar to LUX and BOA in *A. thaliana*, interacting with ELF3 and ELF4 to regulate the circadian clock. ROC40 encodes an MYB transcription factor homologous to CCA1/LHY, the key regulators of the morning phase. ROC66 encodes a CONSTANS-like protein that significantly affects the *C. reinhardtii* clock but has a limited impact in *A. thaliana*.

The mRNA levels of ROC15, ROC40, ROC66, and ROC75 exhibit circadian fluctuations. While the mRNA of ROC114 is expressed continuously throughout the circadian cycle, the level of ROC66 peaks during the mid-subjective night. The mRNA levels of ROC15 and ROC40 oscillate with identical or closely matching phases, peaking during the mid- to late subjective night. ROC75 mRNA levels reach their maximum during the early to mid-subjective day [22].

- **ROC75 gene**—The ROC75 gene is essential for maintaining the normal rhythmicity of bioluminescence rhythms in **continuous light (LL)** conditions. When mutated, the bioluminescence rhythm shows lower amplitude, but it remains rhythmic, unlike in **constant darkness (DD)**, where complete arrhythmicity is observed. This indicates that ROC75 is critical for sustaining the circadian rhythm in the absence of light. However, in the presence of light, ROC75 is not absolutely required, as light itself compensates for its role in rhythm maintenance [23]. Under LL conditions, ROC75 mRNA expression peaks during the early subjective day [22], and its expression is also high under DD conditions. In Light–Dark (LD) conditions, the ROC75 mRNA peaks at dawn and then declines, indicating a downregulation by light [23].
- **ROC15 gene**—To understand the role of ROC15, it is important to understand the **light-induced phase resetting** in *C. reinhardtii*. While circadian clocks can sustain their oscillation independently of external signals, they require "resetting" when there is a discrepancy between their cycles and the environmental daily patterns. Light is a primary signal that serves to reset the circadian clock [33]. In *C. reinhardtii*, light-induced phase resetting of the circadian clock is mediated by ROC15, which directly responds to light and dark signals and likely plays a role in transmitting light/dark information to the circadian clock. ROC15 also encodes a protein resembling a transcription factor with a GARP DNA-binding motif [34]. Upon light exposure, a rapid decline in ROC15 expression occurs, independent of the circadian phase. The correlation between ROC15 degradation and the phase-resetting mechanism highlights its essential role in circadian regulation.

Additionally, phosphorylation of ROC15 occurs both in darkness and upon light exposure. In darkness, phosphorylation is likely related to circadian oscillation, whereas in light, it may facilitate light-induced degradation. While the exact signaling pathways linking light to ROC15 remain unclear, the simultaneous decline of ROC15 after blue and red-light exposure suggests a common underlying mechanism, even if the light signals are transmitted through distinct pathways [7]. ROC15 shows strong light-induced responses to violet (410 nm) and ultraviolet (360 nm) light pulses, but is not affected by far-red-light pulses (710 nm and 760 nm) in the wild type (WT) strain.

- **ROC40 gene**—In *Chlamydomonas*, the expression levels of ROC40 showed similarities to those of CCA1/LHY [34, 35]. The presence of CHLAMY1 within the ROC40 gene [29] suggests that CHLAMY1 could play a role in the circadian clock.
- The genes **ROC15, ROC40, and ROC66** exhibit strong circadian rhythms, peaking at subjective night under LL conditions. This pattern of expression is consistent across DD and LD conditions [23].
- The expression of **ROC59** is strongly influenced by light. ROC75 serves as a repressor, regulating ROC59 expression in response to light, especially during the day. Under DD conditions, ROC59 expression is much lower. In LD conditions, the expression of ROC59 peaks during the day, but this peak is controlled, with ROC75 likely repressing ROC59 expression during the light phase. In LL conditions, ROC59 expression remains high, and ROC75 is involved in regulating this light-dependent expression during the day [23].

There are several other genes in *Chlamydomonas* that influence the circadian period of phototactic rhythms [36, 37].

3.1.4 Plastid Rhythmicity in *C. reinhardtii*

Plastids are essential organelles in plant cells and are believed to have emerged through an endosymbiotic event involving a proto eukaryotic cell and a photosynthetic prokaryote [38, 39]. Rather than developing independently, plastids replicate via binary fission from pre-existing plastids. *C. reinhardtii* features a prominent chloroplast that divides in synchrony with the host cells, making it an excellent model organism for investigating chloroplast division.

In plant plastid division, the proteins FTSZ1 and FTSZ2 (Filamenting temperature-sensitive Z-like proteins 1 and 2) congregate at the center of the plastid to create a Z-ring, akin to the role of FTSZ in bacterial cell division. The positioning of this Z-ring at the plastid's midpoint is vital for successful division, with its placement regulated by MIN proteins: MINC, MIND, and MINE [40–42]. Notably, higher plants lack a MINC homologue, suggesting its loss in evolution, substituted by a non-standard MINC-like protein.

FTSZ, MIND, and MINE1 have also been extracted from *C. reinhardtii*. Both MIND and MINE1 are proteins encoded by the nucleus, meaning they are directed by nuclear genes, but these proteins are likely transported to the chloroplast, aided by N-terminal plastid transit peptides. In *C. reinhardtii*, the mRNA levels of FTSZ and MIN genes exhibit a circadian rhythm, peaking at the transition between light and dark in a light–dark (LD) cycle, just prior to chloroplast division. These rhythmic expression patterns are also noted in both constant dark (DD) and constant light (LL) conditions. The peak in expression closely aligns with the onset of chloroplast division, suggesting that the expression levels of the FTSZ and MIN genes regulate this process [43].

The circadian rhythm of cell division in *C. reinhardtii* is well documented [44]. Numerous other genes are influenced by this rhythm [45–51]. Conversely, the expression of FTSZ, MIND, and MINE1 is directly connected to cell and chloroplast division, but their circadian rhythm persists for only one cycle in LL.

3.1.5 Role of Circadian Component in Photoprotection

In *C. reinhardtii*, light-harvesting complex stress-related protein 3 (LHCSR3) enables energy-dependent quenching (qE) to prevent damage from excessive light [52], which can lead to overexcitation of photosystems, photoinhibition, and the formation of harmful reactive oxygen species.

This green alga has diverse photoreceptors that allow it to sense and adapt to a broad light spectrum. Phototropin (PHOT) and cryptochromes (CRY), known blue-light receptors, play key roles in this process [53, 54]. PHOT regulates qE by inducing LHCSR3 expression under high light [55]. Photolyase Homolog 1 (pCRY), a cryptochrome/photolyase homolog, absorbs UV-A and blue light [56], influencing circadian clock regulation and phase response behavior.

Photoreceptors are crucial for circadian entrainment, period maintenance, and LHCSR3 regulation [57].

A proposed hypothesis suggests that in the morning, red light reduces the need for LHCSR3, possibly through an unidentified red-light photoreceptor activating ROC75 to decrease LHCSR3 expression. As the day progresses and blue light intensifies, the risk of photodamage increases. This triggers pCRY degradation, leading to ROC75 attenuation, which in turn permits LHCSR3 expression. The increased LHCSR3 then dissipates excess energy, preventing damage, and ensuring proper photo-acclimation to changing light conditions throughout the day [57] (Fig. 3).

3.1.6 Circadian-Regulated Behaviors and Metabolism in *Chlamydomonas*

Unicellular algae can move within aquatic environments or damp soil, demonstrating both passive and active forms of movement, where the passive movement is facilitated by ocean currents. For example, in *C. reinhardtii*, a biflagellate alga, the directional movements are influenced by its circadian clock. Notably, green macroalgae of the species *A. linnaei* display a rhythm of buoyancy regulated by circadian cycles when

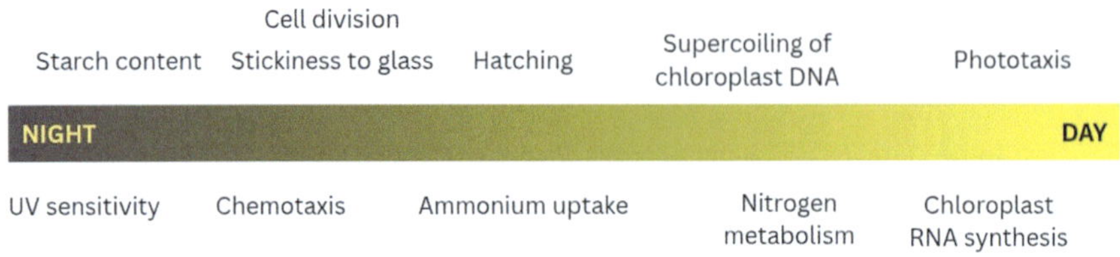

Fig. 3 Diagrammatic representation of the circadian cycle of *Chlamydomonas* influenced by light as its primary zeitgeber. Processes like UV sensitivity, starch content, chemotaxis, stickiness to glass, cell division are active at night time whereas hatching and ammonium uptake are active at transition from night to day. Other activities like supercoiling of chloroplast DNA, nitrogen metabolism, chloroplast RNA synthesis and phototaxis occur during day time

they aggregate into spherical forms. The rhythm of phototaxis, or response to light, in *C. reinhardtii* can be automatically observed and has even been researched in space. This phototactic rhythm is most pronounced during the subjective daytime, while the chemotactic response—movement toward ammonium—peaks at subjective night. Yet, even with this nighttime peak, the algae postpone ammonium uptake until daylight begins. Furthermore, the uptake of nitrite occurs at dawn, aligning with heightened activity of nitrite reductase. These behaviors showcase straightforward but synchronized clock-controlled reactions in unicellular algae. Additional circadian-regulated processes, including cell division, sensitivity to UV light, adhesion to glass, and starch accumulation have also been recorded, highlighting the complex nature of these rhythmic biological activities [10].

Phototaxis and Chemotaxis in *Chlamydomonas*

In *Chlamydomonas*, the ability to move toward light (phototaxis) is robust during daylight hours and diminishes at night. This rhythmic behavior persists in both constant darkness (DD) and constant light (LL), showcasing heightened phototaxis during the subjective day and reduced phototaxis during the subjective night. The cells also exhibit a rhythmic response to ammonium (chemotaxis), which is more pronounced at night compared to daytime. This pattern continues even when the cells are subjected to varying LL conditions. Ammonium absorption, evaluated through light-dependent [^{14}C] methylammonium uptake, displays a circadian rhythm. Notably, the peak in chemotaxis to ammonium does not coincide with the peak in methylammonium uptake. Chemotaxis reaches its maximum during the middle of the subjective night, whereas the highest uptake of methylammonium occurs roughly six hours later, at subjective dawn. Both the ammonium uptake and chemotaxis rhythms are maintained under constant conditions, and the presence of light influences the intensity of the rhythm [58]

Regulation of Nitrate Reduction Enzymes During the *C. Reinhardtii* Cell Cycle

In a 12-h light/dark cycle, the activity of ferredoxin-nitrite reductase in *C. reinhardtii* varies over 24 h, reaching its peak in the middle of the light phase. A similar fluctuation pattern is noted for nitrate reductase activity, but this is not observed for the enzymes that are involved in ammonium assimilation [59]. The regulation of nitrate reduction enzymes seems to be distinct from that of ammonium-assimilatory enzymes such as glutamine synthetase and glutamate synthase, which increase in the light period and maintain high levels in the dark, to manage ammonia generated during protein turnover associated with cell division.

Cell Division

The progression of the cell cycle is mainly regulated by cyclin-dependent kinases (CDKs), which oversee the essential transitions within the cell cycle, encompassing DNA replication, mitosis, and cytokinesis [60]. In 1970, Bruce suggested that the circadian clock directed cell division in *Chlamydomonas*, a proposal later disputed by Spudich and Sager in 1980, who contended that circadian rhythms did not affect cell division in *Chlamydomonas*. However, after 25 years of this, [44] conclusively established that the circadian clock does indeed influence cell division in *Chlamydomonas*. Additionally, it was demonstrated that light intensity significantly affects the resetting of the circadian clock in in *Chlamydomonas*.

3.2 Cyanobacteria

Cyanobacteria, a diverse group of prokaryotic microorganisms, are among the oldest known photosynthetic organisms on Earth. They play a critical role in global carbon and nitrogen cycles and are well-known for their ability to perform oxygenic photosynthesis. Unlike eukaryotic microalgae such as diatoms, cyanobacteria have a relatively simple cellular structure but exhibit complex circadian rhythms that regulate their metabolic and physiological processes. Circadian regulation in cyanobacteria influences key functions such as photosynthesis, nitrogen fixation, and cell division, enabling them to adapt to varying environmental conditions. Numerous cellular activities have been documented to be influenced by circadian clocks in cyanobacteria. For instance, the cell cycle is controlled in both the unicellular *S. elongatus* and in *Anabaena*. Likewise, experimental findings reinforce the idea that the competence state in *S. elongatus* is modulated by the circadian clock [61]. Cyanobacteria are the only prokaryotes known to date to have a fully operational circadian clock [62].

Research on *Synechococcus elongatus* has provided extensive insights into the molecular mechanisms of cyanobacterial circadian clocks. Studies indicate that *S. elongatus* maintains a robust circadian rhythm controlled by a core oscillator comprising three key proteins: KaiA, KaiB, and KaiC. These proteins form a self-sustaining feedback loop that regulates gene expression and metabolic activity in a 24-h cycle [63–65]. This simple yet highly effective circadian system allows *S. elongatus* to optimize its growth and survival in fluctuating light conditions [66] (Fig. 4).

Another significant cyanobacterium *Prochlorococcus marinus* plays a crucial role in the planet's ecology and is likely the most prevalent photosynthetic organism on Earth [67]. *P. marinus* exhibits circadian regulation of photosynthesis and gene expression despite its streamlined genome. Natural populations of *Prochlorococcus* exhibit strong synchronization due to the daily light and dark cycles, showing a diel

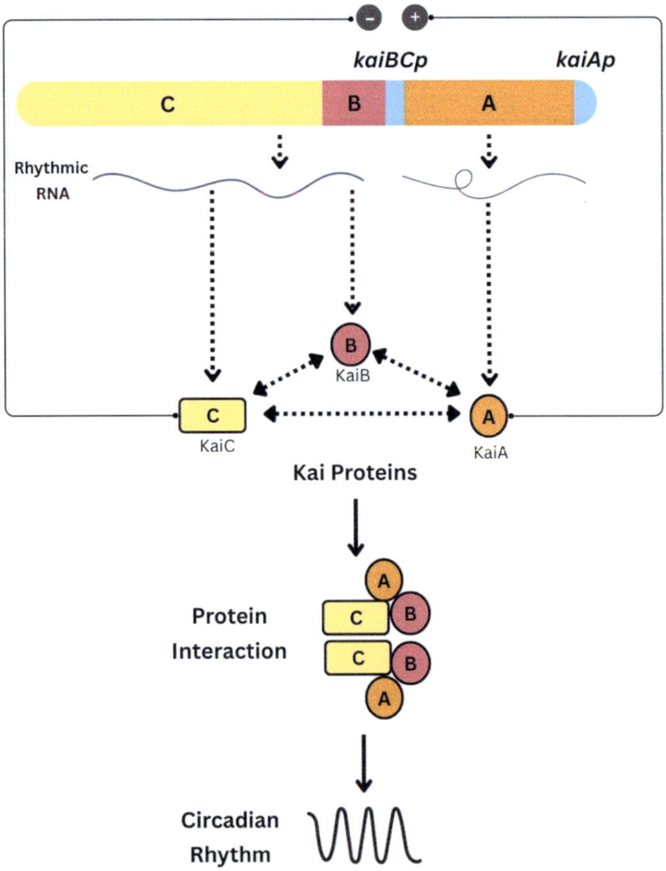

Fig. 4 Model for the circadian clock of *Synechococcus* shows that the kaiA gene is transcribed into a single messenger RNA, whereas the kaiB and kaiC genes are co-transcribed. The promoters for both kaiA and kaiBC exhibit rhythmic activity, leading to fluctuations in the levels of kaiA and kaiBC mRNAs. Proteins KaiA, KaiB, and KaiC interact to create a multimeric complex. The feedback mechanisms involving the proteins KaiA (which provides positive feedback) and KaiC (which exerts negative feedback) on the kaiBC promoter (kaiBCp) are illustrated

pattern in both cell cycling and gene expression [68–70]. Unlike *S. elongatus*, *P. marinus* lacks a complete set of Kai proteins but still demonstrates diel patterns in cellular activity [71]. *Prochlorococcus* exists in a fairly straightforward and stable environment. Additionally, it is absent from high latitude areas and thus does not undergo significant variations in day length. In this type of environment, having a timekeeping mechanism, like an hourglass—meaning an organism that checks the time once daily (most likely at dawn)—would suffice, as there would be no need for a timer that operates beyond 24 h [72].

The role of circadian rhythms in nitrogen fixation has been extensively studied in *Cyanothece sp.*, a unicellular nitrogen-fixing cyanobacterium. Unlike many cyanobacteria that perform photosynthesis and nitrogen fixation in separate cell types, *Cyanothece* carries out both processes within the same cell but in different temporal phases. Circadian regulation ensures that photosynthetic activity occurs during the day while nitrogen fixation takes place at night [73], preventing oxygen-sensitive nitrogenase enzymes from being inactivated by photosynthetically generated oxygen. This separation allows *Cyanothece* to thrive in nitrogen-limited environments.

Nostoc punctiforme, a filamentous cyanobacterium from the Nostocaceae family, when grown continuously under constant light and stable conditions for extended periods, this cyanobacterium maintains a distinct separation in its gene expression between light-dependent and dark-dependent processes. During the day, genes associated with sugar metabolism, peptidoglycan biosynthesis, and motility show increased activity. Conversely, at night, genes responsible for chromosome repair and nucleotide metabolism become more active [74].

In general, cyanobacteria exhibit an impressive ability to adapt through their circadian rhythms, which govern essential processes such as photosynthesis, nitrogen fixation, and gene expression. These rhythms are influenced by mechanisms like the KaiABC protein system found in *S. elongatus*, as well as simpler adaptations in *P. marinus*, allowing these organisms to align their cellular functions with environmental cycles. By temporally organizing their metabolic activities, as demonstrated in *Cyanothece* and *N. punctiforme*, cyanobacteria effectively manage energy requirements and thrive in a variety of ecological environments. Their circadian systems highlight evolutionary creativity, facilitating survival and productivity in ever-changing surroundings.

3.3 Diatoms

Diatoms are a diverse and ecologically significant group of microalgae that thrive in marine and freshwater ecosystems. They, like other organisms that carry out photosynthesis, adapt their biological processes to the cycles of light and darkness throughout the day. Research indicates that they exhibit daily patterns in critical functions such as photosynthesis [75, 76], pigment synthesis [77, 78], cell division [79], and gene expression [80–82]. Investigations conducted in both laboratory settings and natural habitats have also revealed that these daily cycles influence the growth of diatoms [83, 84] and the expression of genes linked to photosynthesis [85, 86].

Another diatom, *Skeletonema costatum*, is a species that can be found throughout the ocean. This organism frequently creates significant blooms that impact both biogeochemical cycles and the food web. Cells of *S. costatum* can adjust their circadian rhythm by increasing the expression of genes related to the circadian clock [87]. Three phytochrome genes (PHYC, PHYD, and PHYE), alongside a gene involved in circadian regulation (CRY1), as well as zeitlupe (ZTL), flarin binding, kelch repeat, f-box 1 (FKF1), and LOV kelch protein 2 (LKP2), which are all part of the circadian clock's input pathway, can be found in both *S. costatum* and *P. tricornutum*. However, the phytochrome genes PHYA and PHYB are exclusive to *S. costatum*. Three essential genes of the core oscillator of the circadian clock (CCA1, LHY, and TOC1) can be found in both diatom species. The levels of three essential clock genes (CCA1, LHY, and TOC1) in *S. costatum* and *P. tricornutum* exhibited distinct oscillations over a 24-h period [88].

To elaborate, a molecular model representing the circadian clock in marine diatoms has been suggested. The input pathway consists of five genes responsible for red-light receptors (PHYA, PHYB, PHYC, PHYD, and PHYE), one gene for a blue-light receptor (CRY1), and three genes that regulate the clock downstream (ZTL, LKP2, and FKF1), which relay environmental signals to the core oscillator. The core oscillator comprises nine genes that create a complex feedback network, including a core loop featuring CCA1/LHY and TOC1, a morning loop involving CCA1/LHY along with PRR7/PRR9, and an evening loop managed by TOC1. These gene interactions yield signals that oscillate periodically over a 24-h cycle. The only gene noted in the output pathway was CAB2 [88].

In summary, marine diatoms are greatly affected by abiotic factors [89], with light being the most crucial environmental factor that influences both cellular growth and the circadian rhythm [90]. Although most essential clock genes are shared among the two diatom species examined, there are some variations present. This indicates that marine diatoms have developed circadian clock mechanisms that are specifically tailored to their distinct environmental conditions, which helps them manage stress more effectively and regulate their biological functions in relation to light and other influences [88].

3.4 Dinoflagellates Display Circadian Rhythms

Dinoflagellates exhibit circadian rhythms in swimming behavior (aggregation) that differ from their bioluminescence rhythms. Under specific conditions, the free-running periods and phase responses to light pulses vary between these two outputs, implying the presence of two distinct oscillators within single cells. The aggregation oscillator is sensitive to both blue and red light, while the bioluminescence oscillator responds predominantly to blue light. Increasing blue-light intensity shortens the bioluminescence period, whereas increasing red light lengthens it—suggesting that different photoreceptors mediate light input to each oscillator [11].

3.5 *Rhythms in Other Algal Species*

3.5.1 Green Algae

Picochlorum celeri is a resilient chlorophyte microalga known for its exceptional tolerance to high light and temperature conditions. Among key circadian regulators, the *CCA1* gene shows distinct expression patterns. The variation indicates potential circadian modifications that enable *P. celeri* to thrive in fluctuating light environments [91]. This divergence in *CCA1* regulation aligns with the overall reduction of circadian-related genes in *P. celeri* compared to higher plants like *Arabidopsis thaliana*, highlighting its streamlined but efficient regulatory system. At the molecular level, the translated protein sequence of *P. celeri CCA1* exhibits strong similarity to *A. thaliana* homologs of *CCA1* and *LHY*, featuring a conserved domain associated with the SANT family of MYB-like DNA-binding sequences. Hereafter, this protein is referred to as *Picochlorum CCA1* [91].

The smallest single-celled eukaryote, *Ostreococcus tauri,* a single-celled green alga belonging to the Mamiellophyceae class, has the smallest known eukaryotic genome among photosynthetic organisms. The circadian rhythm of *O. tauri* is considered a simplified version of the plant model *Arabidopsis thaliana* and provides valuable insights into minimalistic clock systems [23]. Research indicates that the circadian clock in *O. tauri* operates through a feedback loop involving both transcriptional and translational processes with essential clock genes. This regulation aligns these genes with external light cycles, enhancing photosynthesis and metabolic functions. Most differentially expressed genes in *O. tauri* cells subjected to a 12 h/12 h light–dark cycle are transcriptionally regulated during the transition from light to dark, particularly those associated with cell cycle, DNA replication, nuclear transcription, and photosynthesis [60, 92]. The rhythmic expression of genes in *O. tauri* affects several biological processes, including carbon metabolism and cellular division. It demonstrates a resilient yet streamlined circadian mechanism, establishing it as a crucial model for investigating fundamental circadian regulation in photosynthetic eukaryotes. Specific clock-associated genes in *O. tauri* include homologs of *CCA1/LHY* and *TOC1*, which exhibit distinct expression patterns in response to light–dark cycles. Other Circadian clock proteins include—CO, CRY1, CK1 and GSK3. Additionally, *O. tauri*'s clock interacts with metabolic pathways, ensuring synchronization between energy production and cellular division cycles. TOC1 expression peaks at subjective dusk, like its counterpart in *Arabidopsis thaliana.* And when CCA1 levels rise, TOC1 expression decreases [11]. The CCA1 gene shows peak expression around dawn when exposed to daily light cycles [91]. The streamlined clock network of *O. tauri* contrasts with more complex plant circadian mechanisms, highlighting its evolutionary adaptation to unicellular life forms.

Platymonas helgolandica var. Tsingtaoensis (commonly referred to as *P. helgolandica*) is a marine microalga capable of accumulating high starch content, particularly amylose, under specific circadian rhythms. The circadian clock in *P. helgolandica* operates through an intricate feedback loop involving multiple transcriptional regulators. Key clock components include:

- Timing of cab expression 1 (toc1)—Expressed at the start of the night, influencing nocturnal processes.
- Late Elongated Hypocotyl (lhy) and Circadian Clock-Associated 1 (cca1)—MYB transcription factors that regulate diurnal activity by repressing each other.
- GIGANTEA (gi) and clock-associated PAS protein ZTL (ztl)—Regulate *toc1*, impacting nighttime activity.
- Pseudoresponse regulators prr7 and prr9—Modulate the interplay between lhy, cca1, and daytime cellular functions.

This coordinated regulation ensures the synchronization of metabolic activities, such as starch biosynthesis. The ability of *P. helgolandica* to modulate starch accumulation according to its internal clock and environmental cues highlights the significance of circadian rhythms in microalgal biotechnology and bioenergy applications [93].

Chromera velia is a free living marine alga and the closest known photosynthetic relative of apicomplexan parasites. The life cycle of *C. velia* includes the formation of motile zoospores, which is regulated by light and follows a periodicity of approximately 24 h. The release of zoospores occurs rhythmically, beginning at dawn, peaking at midday, and declining before dusk. Under continuous darkness, this rhythmic release persists for only three cycles, while under constant light conditions, zoospore release remains robust but loses its cyclic nature. This suggests the presence of an internal circadian clock in *C. velia*. Two candidate myb-like proteins, CCA1 and LHY, have been identified as crucial components for clock function, highlighting similarities to circadian regulators in plants and algae. The circadian cycle of *C. velia* is largely regulated at the translational level, resembling the clock mechanisms observed in dinoflagellates [94].

3.5.2 Red Algae

Cyanidioschyzon merolae is a unicellular red alga with a highly streamlined genome, making it an ideal model for studying circadian rhythms. Photosynthetic and respiratory activities in *C. merolae* peak during the morning and gradually decline toward the evening and night. In contrast, pathways involved in anaerobic pyruvate metabolism are upregulated at night, indicating a temporal segregation of metabolic processes to optimize energy production [95]. The circadian cycle of *C. merolae* exhibits a distinct regulation of gene expression, with chromosome replication genes peaking at hour 11 of the light period. Additionally, genes involved in fermentation pathways, such as *PFL1, ADH, PAT1/2, ACK1/2*, and *LDH*, show upregulation during the evening, like *C. reinhardtii*. However, unlike *C. reinhardtii*, pyruvate dehydrogenase components in *C. merolae* are downregulated in the evening [95]. A genome-wide analysis of *C. merolae* revealed diel transcriptomic changes in 1,979 out of 4,775 nucleus-encoded genes, with 454 of these genes exhibiting periodic expression patterns dependent on cell cycle progression. Genes involved in metabolic pathways peak during the day, whereas those associated with chromosome replication and cell division peak at night under 12-h light/dark (LD) cycles [96].

In *Neopyropia yezoensis*, genes associated with carbon–nitrogen assimilation, polyunsaturated fatty acid synthesis, and starch metabolism peak in the morning, while those linked to the pentose phosphate pathway, cell cycle regulation, and DNA replication are upregulated in the evening. This segregation suggests a protective mechanism to prevent light-induced DNA damage. Key cell cycle regulators such as *NyCYCB* (Cyclin B) and *CDC20* exhibit peak expression at dusk, ensuring tight control of mitotic progression. Similarly, genes encoding cryptochrome/photolyase family proteins and histone modification proteins play crucial roles in regulating diurnal gene expression [97].

Genome-wide expression profiling of *Gracilariopsis chorda* identified 3098 rhythmic genes, with specific cryptochrome genes exhibiting distinct diurnal expression patterns [98].

The study of circadian rhythms in red algae provides critical insights into metabolic optimization, cell cycle regulation, and adaptive strategies under diel cycles. The distinct temporal segregation of metabolic and cell cycle processes in *C. merolae*, *N. yezoensis*, and *G. chorda* highlights the importance of circadian clocks in marine and terrestrial environments. Understanding these mechanisms may further applications in biotechnology, aquaculture, and biofuel production [95–98].

4 Conclusion

This chapter highlights the fundamental role of circadian rhythms in algal life. From regulating buoyancy and movement in green algae, controlling cell division, UV light sensitivity, and starch accumulation, these rhythms enhance algal survival. At the cellular level, they govern key functions such as transcription and DNA repair within the nucleus. Their ability to regulate diverse processes stems from their responsiveness to environmental cues, which serve as inputs processed by the central oscillator and executed through gene expression.

One striking example is the seasonal green tide dominated by *U. prolifera*, where rapid growth and reproduction drive blooms that appear and disappear between spring and autumn. Environmental stressors such as high light, temperature, grazing, damage, and eutrophication shift biomass allocation toward reproduction, triggering sporogenesis and forming a spore bank that fuels future blooms. Beyond daily 24-h rhythms, some species also exhibit seasonal, or circannual, rhythms, as seen in *A. tamarense* (formerly *Gonyaulax*), a key unicellular model for studying these patterns.

Circadian rhythms trace back to cyanobacteria, the earliest known prokaryotes with a circadian clock. Core clock components, including PRR and RVE genes, are conserved across the green lineage. In angiosperms, the Evening Complex (EC), regulated by ELF3 and ELF4, plays a pivotal role, and its conservation in streptophytes highlights evolutionary retention.

Among the most valuable models for studying circadian rhythms is *C. reinhardtii*, owing to its well-regulated traits such as phototaxis, chemotaxis, and cell adhesion. Its simple life cycle, ease of mutant isolation, and expanding genetic tools further enhance its research utility. Comparative studies with other algae have identified key circadian genes in *Chlamydomonas*, particularly the ROC protein family (ROC15, ROC40, ROC55, ROC66, ROC75, and ROC114). Notably, *C. reinhardtii* shares homologous genes with *Arabidopsis thaliana*, making it a crucial system for circadian studies. For instance, ROC15 and ROC75 encode GARP transcription factors akin to LUX (PCL1) and BOA (NOX) in *A. thaliana*. These proteins interact with ELF3 and ELF4 to regulate the circadian clock by promoting *CCA1* expression. However, *C. reinhardtii* also possesses circadian clock-related genes that lack homologs in *A. thaliana*, emphasizing its unique contribution to circadian biology.

Beyond *C. reinhardtii*, various other model organisms have expanded our understanding of circadian rhythms. *C. merolae*, a unicellular red alga with a highly streamlined genome, offers a simplified system for studying cellular and genetic regulation. Similarly, the circadian rhythm of *O. tauri* is considered a minimalistic version of the plant model *Arabidopsis thaliana*, providing kcy insights into simplified clock mechanisms. Other valuable models include cyanobacteria (*S. elongatus*), diatoms (*S. costatum*), and dinoflagellates (*Gonyaulax polyedra*), each contributing to a broader understanding of biological clocks across diverse evolutionary lineages.

References

1. Lincoln G (2019) A brief history of circannual time. J Neuroendocrinology 31
2. Helm B, Lincoln GA (2017) Circannual rhythms anticipate the earth's annual periodicity. Biol Timekeep: Clocks, Rhythm Behav, pp 545–569
3. Kalita TL, Titlyanov EA (2013) Influence of temperature on the infradian growth rhythm in *Ulva lactuca* (Chlorophyta). Eur J Phycol 48:210–220
4. Crosthwaite SK, Loros JJ, Dunlap JC (1995) Light-induced resetting of a circadian clock is mediated by a rapid increase in frequency transcript. Cell 81:1003–1012
5. Dvornyk V (2016) Evolution of the circadian clock system in cyanobacteria: a genomic perspective. Inter J Algae 18:5–20
6. Devlin PF (2002) Signs of the time: environmental input to the circadian clock. J Exp Bot 53:1535–1550
7. Niwa Y, Matsuo T, Onai K, Kato D, Tachikawa M, Ishiura M (2013) Phase-resetting mechanism of the circadian clock in *Chlamydomonas reinhardtii*. Proc Natl Acad Sci USA 110:13666–13671
8. Roenneberg T, Mittag M (1996) The circadian program of algae. Semin Cell Dev Biol 7:753–763
9. Häfker NS, Andreatta G, Manzotti A, Falciatore A, Raible F, Tessmar-Raible K (2023) Rhythms and clocks in marine organisms. Annu Rev Mar Sci 15:509–538
10. Petersen J, Rredhi A, Szyttenholm J, Mittag M (2022) Evolution of circadian clocks along the green lineage. Plant Physiol 190:924–937

11. Noordally ZB, Millar AJ (2014) Clocks in algae. Biochemistry 54:171–183
12. Jiang J, Chen Y, Zhang R, Zhu W, Liu F, Xu N, Li Y (2025) New insights on the impact of light, photoperiod and temperature on the reproduction of green algae *Ulva prolifera* via transcriptomics and physiological analyses. Mar Pollut Bull 211:117393
13. Serrano-Bueno G, Romero-Campero FJ, Lucas-Reina E, Romero JM, Valverde F (2017) Evolution of photoperiod sensing in plants and algae. Curr Opin Plant Biol 37:10–17
14. Mittag M (1996) Conserved circadian elements in phylogenetically diverse algae. Proc Natl Acad Sci USA 93:14401–14404
15. Chaix A, Zarrinpar A, Panda S (2016) The circadian coordination of cell biology. J Cell Biol 215:15–25
16. Schulze T, Prager K, Dathe H, Kelm J, Kießling P, Mittag M (2010) How the green alga *Chlamydomonas reinhardtii* keeps time. Protoplasma 244:3–14
17. Nikaido SS, Johnson CH (2000) Daily and circadian variation in survival from ultraviolet radiation in *Chlamydomonas reinhardtii*. Photochem Photobiol 71:758
18. Mittag M, Kiaulehn S, Johnson CH (2005) The circadian clock in *Chlamydomonas reinhardtii*. What is it for? What is it similar to? Plant Physiol 137:399–409
19. Harris EH (2001) CHLAMYDOMONAS as a model organism. Annu Rev Plant Physiol Plant Mol Biol 52:363–406
20. Grossman AR, Harris EE, Hauser C, Lefebvre PA, Martinez D, Rokhsar D, Shrager J, Silflow CD, Stern D, Vallon O, Zhang Z (2003) *Chlamydomonas reinhardtii* at the crossroads of genomics. Eukaryot Cell 2:1137–1150
21. Young MW, Kay SA (2001) Time zones: a comparative genetics of circadian clocks. Nat Rev Genet 2:702–715
22. Matsuo T, Okamoto K, Onai K, Niwa Y, Shimogawara K, Ishiura M (2008) A systematic forward genetic analysis identified components of the Chlamydomonas circadian system. Genes Dev 22:918–930
23. Matsuo T, Iida T, Ohmura A, Gururaj M, Kato D, Mutoh R, Ihara K, Ishiura M (2020) The role of ROC75 as a daytime component of the circadian oscillator in *Chlamydomonas reinhardtii*. PLoS Genet 16:e1008814
24. Robson F, Costa MMR, Hepworth SR, Vizir I, Pin˜eiro M, Reeves PH, Putterill J, Coupland G (2001) Functional importance of conserved domains in the flowering-time gene CONSTANS demonstrated by analysis of mutant alleles and transgenic plants. Plant J 28:619–631
25. Nusinow DA, Helfer A, Hamilton EE, King JJ, Imaizumi T, Schultz TF, Farré EM, Kay SA (2011) The ELF4–ELF3–LUX complex links the circadian clock to diurnal control of hypocotyl growth. Nature 475:398–402
26. Ledger S, Strayer C, Ashton F, Kay SA, Putterill J (2001) Analysis of the function of two circadian-regulated CONSTANS-LIKE genes. Plant J 26:15–22
27. Matsuo T, Ishiura M (2010) New Insights into the circadian clock in chlamydomonas. Int Rev Cell Mol Biol, pp 281–314
28. Matsuo T, Ishiura M (2011) *Chlamydomonas reinhardtii* as a new model system for studying the molecular basis of the circadian clock. FEBS Lett 585:1495–1502
29. Iliev D, Voytsekh O, Schmidt E-M, Fiedler M, Nykytenko A, Mittag M (2006) A Heteromeric RNA-binding protein is involved in maintaining acrophase and period of the circadian clock. Plant Physiol 142:797–806
30. Dathe H, Prager K, Mittag M (2012) Novel interaction of two clock-relevant RNA-binding proteins C3 and XRN1 in *Chlamydomonas reinhardtii*. FEBS Lett 586:3969–3973
31. Matsuo T, Iida T, Ishiura M (2012) N-terminal acetyltransferase 3 gene is essential for robust circadian rhythm of bioluminescence reporter in *Chlamydomonas reinhardtii*. Biochem Biophys Res Commun 418:342–346
32. Müller N, Wenzel S, Zou Y, Künzel S, Sasso S, Weiß D, Prager K, Grossman A, Kottke T, Mittag M (2017) A plant cryptochrome controls key featurs of the chlamydomonas circadian clock and its life cycle. Plant Physiol 174:185–201
33. Kinoshita A, Niwa Y, Onai K, Yamano T, Fukuzawa H, Ishiura M, Matsuo T (2017) CSL encodes a leucine-rich-repeat protein implicated in red/violet light signaling to the circadian clock in Chlamydomonas. PLoS Genet 13:e1006645

34. Wang ZY, Kenigsbuch D, Sun L, Harel E, Ong MS, Tobin EM (1997) A Myb-related transcription factor is involved in the phytochrome regulation of an Arabidopsis Lhcb gene. Plant Cell 9:491–507
35. Schaffer R, Ramsay N, Samach A, Corden S, Putterill J, Carré IA, Coupland G (1998) The late elongated hypocotyl mutation of Arabidopsis disrupts circadian rhythms and the photoperiodic control of flowering. Cell 93:1219–1229
36. Jeong B, Wu-Scharf D, Zhang C, Cerutti H (2002) Suppressors of transcriptional transgenic silencing in Chlamydomonas are sensitive to DNA-damaging agents and reactivate transposable elements. Proc Natl Acad Sci USA 99:1076–1081
37. Zhou B-BS, Elledge SJ (2000) The DNA damage response: putting checkpoints in perspective. Nature 408:433–439
38. Gray MW (1999) Evolution of organellar genomes. Curr Opin Genet Dev 9:678–687
39. McFadden GI (2001) Chloroplast origin and integration. Plant Physiol 125:50–53
40. Colletti KS, Tattersall EA, Pyke KA, Froelich JE, Stokes KD, Osteryoung KW (2000) A homologue of the bacterial cell division site-determining factor MinD mediates placement of the chloroplast division apparatus. Curr Biol 10:507–516
41. Maple J, Chua N, Møller SG (2002) The topological specificity factor AtMinE1 is essential for correct plastid division site placement in Arabidopsis. Plant J 31:269–277
42. Reddy S, Dinkins R, Collins G (2002) Overexpression of the Arabidopsis thaliana MinE1 bacterial division inhibitor homologue gene alters chloroplast size and morphology in transgenic Arabidopsis and tobacco plants. Planta 215:167–176
43. Hu Y, Chen Z-W, Liu W-Z, Liu X-L, He Y-K (2008) Chloroplast division is regulated by the circadian expression of FTSZ and MIN genes in *Chlamydomonas reinhardtii*. Eur J Phycol 43:207–215
44. Goto K, Johnson CH (1995) Is the cell division cycle gated by a circadian clock? The case of *Chlamydomonas reinhardtii*. J Cell Biol 129:1061–1069
45. Leu S, White D, Michaels A (1990) Cell cycle-dependent transcriptional and post-transcriptional regulation of chloroplast gene expression in *Chlamydomonas reinhardtii*. Biochimica et Biophysica Acta (BBA)–Gene Structure and Expression 1049:311–317
46. Hwang S, Herrin DL (1994) Control of LHC gene transcription by the circadian clock in *Chlamydomonas reinhardtii*. Plant Mol Biol 26:557–569
47. Jacobshagen S, Johnson CH (1994) Circadian rhythms of gene expression in *Chlamydomonas reinhardtii*: circadian cycling of mRNA abundances of cab II, and possibly of beta-tubulin and cytochrome c. Eur J Cell Biol 64(1):142–152
48. Fujiwara S, Ishida N, Tsuzuki M (1996) Circadian expression of the carbonic anhydrase gene, Cah1, in *Chlamydomonas reinhardtii*. Plant Mol Biol 32:745–749
49. Hwang S, Kawazoe R, Herrin DL (1996) Transcription of tufA and other chloroplast-encoded genes is controlled by a circadian clock in Chlamydomonas. Proc Natl Acad Sci USA 93:996–1000
50. Jacobshagen S, Kindle KL, Johnson CH (1996) Transcription ofCABII is regulated by the biological clock in *Chlamydomonas reinhardtii*. Plant Mol Biol 31:1173–1184
51. Lemaire SD, Stein M, Issakidis-Bourguet E, Keryer E, Benoit V, Pineau B, Gérard-Hirne C, Miginiac-Maslow M, Jacquot J-P (1999) The complex regulation of ferredoxin/thioredoxin-related genes by light and the circadian clock. Planta 209:221–229
52. Peers G, Truong TB, Ostendorf E, Busch A, Elrad D, Grossman AR, Hippler M, Niyogi KK (2009) An ancient light-harvesting protein is critical for the regulation of algal photosynthesis. Nature 462:518–521
53. Huang K, Beck CF (2003) Phototropin is the blue-light receptor that controls multiple steps in the sexual life cycle of the green alga *Chlamydomonas reinhardtii*. Proc Natl Acad Sci USA 100:6269–6274
54. Beel B, Prager K, Spexard M, Sasso S, Weiss D, Müller N, Heinnickel M, Dewez D, Ikoma D, Grossman AR, Kottke T, Mittag M (2012) A flavin binding cryptochrome photoreceptor responds to both blue and red light in *Chlamydomonas reinhardtii*. Plant Cell 24:2992–3008

55. Petroutsos D, Tokutsu R, Maruyama S, Flori S, Greiner A, Magneschi L, Cusant L, Kottke T, Mittag M, Hegemann P, Finazzi G, Minagawa J (2016) A blue-light photoreceptor mediates the feedback regulation of photosynthesis. Nature 537:563–566
56. Reisdorph NA, Small GD (2004) The CPH1 gene of *Chlamydomonas reinhardtii* encodes two forms of cryptochrome whose levels are controlled by light-induced proteolysis. Plant Physiol 134:1546–1554
57. Kamrani YY, Matsuo T, Mittag M, Minagawa J (2018) ROC75 is an Attenuator for the circadian clock that controls LHCSR3 expression. Plant Cell Physiol
58. Byrne TE, Wells MR, Johnson CH (1992) Circadian rhythms of chemotaxis to Ammonium and of methylammonium uptake in chlamydomonas. Plant Physiol 98:879–886
59. Pajuelo E, Pajuelo P, Clemente MT, Márquez AJ (1995) Regulation of the expression of ferredoxin-nitrite reductase in synchronous cultures of *Chlamydomonas reinhardtii*. Biochimica et Biophysica Acta (BBA)–Protein Structure and Molecular Enzymology 1249:72–78
60. Moulager M, Monnier A, Jesson B, Bouvet R, Mosser J, Schwartz C, Garnier L, Corellou F, Bouget F-Y (2007) Light-dependent regulation of cell division inostreococcus: evidence for a major transcriptional input. Plant Physiol 144:1360–1369
61. Arbel-Goren R, Dassa B, Zhitnitsky A, Valladares A, Herrero A, Flores E, Stavans J (2024) Spatio-temporal coherence of circadian clocks and temporal control of differentiation in Anabaena filaments. mSystems 9
62. Holtzendorff J, Partensky F, Mella D, Lennon J-F, Hess WR, Garczarek L (2008) Genome streamlining results in loss of robustness of the circadian clock in the marine cyanobacterium *Prochlorococcus marinus* PCC 9511. J Biol Rhythms 23:187–199
63. Iwasaki H, Nishiwaki T, Kitayama Y, Nakajima M, Kondo T (2002) KaiA-stimulated KaiC phosphorylation in circadian timing loops in cyanobacteria. Proc Natl Acad Sci USA 99:15788–15793
64. Rust MJ, Markson JS, Lane WS, Fisher DS, O'Shea EK (2007) Ordered phosphorylation governs oscillation of a three-protein circadian clock. Science 318:809–812
65. Qin X, Byrne M, Mori T, Zou P, Williams DR, Mchaourab H, Johnson CH (2010) Intermolecular associations determine the dynamics of the circadian KaiABC oscillator. Proc Natl Acad Sci USA 107:14805–14810
66. Xu Y, Weyman PD, Umetani M, Xiong J, Qin X, Xu Q, Iwasaki H, Johnson CH (2013) Circadian Yin-Yang regulation and its manipulation to globally reprogram gene expression. Curr Biol 23:2365–2374
67. Partensky F, Hess WR, Vaulot D (1999) Prochlorococcus, a marine photosynthetic prokaryote of global significance. Microbiol Mol Biol Rev 63:106–127
68. Garczarek L, Partensky F, Irlbacher H, Holtzendorff J, Babin M, Mary I, Thomas JC, Hess WR (2001) Differential expression of antenna and core genes in Prochlorococcus PCC 9511 (Oxyphotobacteria) grown under a modulated light–dark cycle. Environ Microbio 3:168–175
69. Holtzendorff J, Partensky F, Jacquet S, Bruyant F, Marie D, Garczarek L, Mary I, Vaulot D, Hess WR (2001) Diel expression of cell cycle-related genes in synchronized cultures of Prochlorococcus sp. Strain PCC 9511. J Bacteriol 183:915–920
70. Holtzendorff J, Marie D, Post AF, Partensky F, Rivlin A, Hess WR (2002) Synchronized expression of ftsZ in natural Prochlorococcus populations of the Red Sea. Environ Microbiol 4:644–653
71. Rocap G, Larimer FW, Lamerdin J, Malfatti S, Chain P, Ahlgren NA, Arellano A, Coleman M, Hauser L, Hess WR, Johnson ZI, Land M, Lindell D, Post AF, Regala W, Shah M, Shaw SL, Steglich C, Sullivan MB, Ting CS, Tolonen A, Webb EA, Zinser ER, Chisholm SW (2003) Genome divergence in two Prochlorococcus ecotypes reflects oceanic niche differentiation. Nature 424:1042–1047
72. Mullineaux CW, Stanewsky R (2009) The rolex and the hourglass: a simplified circadian clock in prochlorococcus? J Bacteriol 191:5333–5335
73. Vinh NX, Chetty M, Coppel R, Gaudana S, Wangikar PP (2013) A model of the circadian clock in the cyanobacterium Cyanothece sp. ATCC 51142. BMC Bioinform 14

74. Kennedy SJ, Risser DD, Paul BG (2024) Diel expression dynamics in filamentous cyanobacteria. Cold Spring Harb Lab
75. Post AF, Dubinsky Z, Wyman K, Falkowski PG (1984) Kinetics of light-intensity adaptation in a marine planktonic diatom. Mar Biol 83:231–238
76. Jallet D, Caballero MA, Gallina AA, Youngblood M, Peers G (2016) Photosynthetic physiology and biomass partitioning in the model diatom Phaeodactylum tricornutum grown in a sinusoidal light regime. Algal Res 18:51–60
77. Owens TG, Falkowski PG, Whitledge TE (1980) Diel periodicity in cellular chlorophyll content in marine diatoms. Mar Biol 59:71–77
78. Hunsperger HM, Ford CJ, Miller JS, Cattolico RA (2016) Differential regulation of duplicate light-dependent Protochlorophyllide oxidoreductases in the diatom Phaeodactylum tricornutum. PLoS ONE 11:e0158614
79. Nelson DM, Brand LE (1979) Cell division periodicity in 13 species of marine phytoplankton on a light: dark cycle1. J Phycol 15:67–75
80. Chauton MS, Winge P, Brembu T, Vadstein O, Bones AM (2012) Gene regulation of carbon fixation, storage, and utilization in the Diatomphaeodactylum Tricornutumacclimated to light/dark cycles. Plant Physiol 161:1034–1048
81. Smith SR, Gillard JTF, Kustka AB, McCrow JP, Badger JH, Zheng H, New AM, Dupont CL, Obata T, Fernie AR, Allen AE (2016) Transcriptional orchestration of the global cellular response of a model pennate diatom to diel light cycling under iron limitation. PLoS Genet 12:e1006490
82. Bilcke G, Osuna-Cruz CM, Santana Silva M, Poulsen N, D'hondt S, Bulankova P, Vyverman W, De Veylder L, Vandepoele K (2021) Diurnal transcript profiling of the diatom Seminavis robusta reveals adaptations to a benthic lifestyle. Plant J 107:315–336
83. Brand LE (1982) Persistent diel rhythms in the chlorophyll fluorescence of marine phytoplankton species. Mar Biol 69:253–262
84. Kheireddine M, Antoine D (2014) Diel variability of the beam attenuation and backscattering coefficients in the northwestern Mediterranean Sea (BOUSSOLE site). JGR Oceans 119:5465–5482
85. Kolody BC, McCrow JP, Allen LZ, Aylward FO, Fontanez KM, Moustafa A, Moniruzzaman M, Chavez FP, Scholin CA, Allen EE, Worden AZ, Delong EF, Allen AE (2019) Diel transcriptional response of a California current plankton microbiome to light, low iron, and enduring viral infection. ISME J 13:2817–2833
86. Coesel SN, Durham BP, Groussman RD, Hu SK, Caron DA, Morales RL, Ribalet F, Armbrust EV (2021) Diel transcriptional oscillations of light-sensitive regulatory elements in open-ocean eukaryotic plankton communities. Proc Natl Acad Sci U.S.A. 118
87. Zhang S-F, Yuan C-J, Chen Y, Chen X-H, Li D-X, Liu J-L, Lin L, Wang D-Z (2016) Comparative transcriptomic analysis reveals novel insights into the adaptive response of *Skeletonema costatum* to changing ambient phosphorus. Front Microbiol 07
88. Zhang S, Wu Y, Lin L, Wang D (2022) Molecular insights into the circadian clock in marine diatoms. Acta Oceanol Sin 41:87–98
89. Irigoien X, Huisman J, Harris RP (2004) Global biodiversity patterns of marine phytoplankton and zooplankton. Nature 429:863–867
90. McClung CR (2001) Circadian rhythms in plants. Annu Rev Plant Physiol Plant Mol Biol 52:139–162
91. Steichen S, Deshpande A, Mosey M, Loob J, Douchi D, Knoshaug EP, Brown S, Nielsen R, Weissman J, Carrillo LR, Laurens LML (2024) Central transcriptional regulator controls photosynthetic growth and carbon storage in response to high light. Nat Commun 15
92. Monnier A, Liverani S, Bouvet R, Jesson B, Smith JQ, Mosser J, Corellou F, Bouget F-Y (2010) Orchestrated transcription of biological processes in the marine picoeukaryote Ostreococcus exposed to light/dark cycles. BMC Genomics 11
93. Shi Q, Zhou Z, Hong Z, Yang Z, Gao Z, Sun L, Fan J (2023) Transcriptomics reveal how circadian regulation contributes to starch hyperaccumulation in marine alga Tetraselmis helgolandica. GCB Bioenergy 15:1477–1493

94. Richtová J, Bazalová O, Horák A, Tomčala A, Gonepogu VG, Oborník M, Doležel D (2023) Circadian rhythms and circadian clock gene homologs of complex alga *Chromera velia*. Front Plant Sci 14
95. Miyagishima S, Era A, Hasunuma T, Matsuda M, Hirooka S, Sumiya N, Kondo A, Fujiwara T (2019) Day/night separation of oxygenic energy metabolism and nuclear DNA replication in the unicellular red alga *Cyanidioschyzon merolae*. mBio 10
96. Fujiwara T, Hirooka S, Ohbayashi R, Onuma R, Miyagishima S (2020) Relationship between cell cycle and diel transcriptomic changes in metabolism in a unicellular red alga. Plant Physiol 183:1484–1501
97. Kominami S, Mizuta H, Uji T (2022) Transcriptome profiling in the marine red alga *Neopyropia yezoensis* under light/dark cycle. Mar Biotechnol 24:393–407
98. Lee J, Yang JH, Weber APM, Bhattacharya D, Kim W-Y, Yoon HS (2024) Diurnal rhythms in the red seaweed gracilariopsis chorda are characterized by unique regulatory networks of carbon metabolism. Mol Biol Evol 41

Chapter 6
Rhythms of Leaf and Stomatal Movements

Prasann Kumar, Shipa Rani Dey, Joginder Singh, and Neha Gogia

1 Introduction

Although plants are rooted in place and cannot move like animals, they are capable of performing various essential movements that are crucial for their survival, adaptation, and productivity. These include nastic (non-directional), tropic (directional), and oscillatory movements, which are triggered by both internal biological signals and external environmental cues such as light, temperature, and humidity. These responses allow plants to optimize the capture of resources like sunlight and water, and to adjust their growth in real time according to changing conditions [17]. Among the most fascinating and ecologically important are rhythmic movements, such as the opening and closing of leaves or stomata. Behaviors like flowers closing at night are not random but part of a complex, coordinated set of physiological responses that demonstrate a plant's ability to sense and dynamically adapt to its environment. These rhythmic patterns offer valuable insights into the oscillatory behavior of leaves and stomata in plant physiology, especially in the context of daily and seasonal environmental changes. Leaf and stomatal rhythms are therefore a vital and intricate aspect of plant behavior, playing a significant role in regulating photosynthesis, gas exchange, water-use efficiency, and overall plant productivity [13, 21].

P. Kumar (✉)
Department of Agronomy, School of Agriculture, Lovely Professional University, Phagwara, Punjab, India
e-mail: prasann0659@gmail.com

S. R. Dey
Department of Agronomy, Faculty of Agricultural Sciences, GLA University, Mathura, Chaumuhan, Uttar Pradesh, India

J. Singh
Department of Botany, Nagaland University, Nagaland, India

N. Gogia
New Haven, CT, USA

A. Kothari Chhajer and V. Singh Gour (eds.), *Introduction to Rhythms in Plants*, Studies in Rhythm Engineering, https://doi.org/10.1007/978-981-95-3592-7_6

Leaf movement is a rhythmic process controlled by the plant's internal biological clock, which follows a 24-h cycle. This natural rhythm is known as a circadian rhythm. One of the most common types of leaf movement is nyctinasty, seen in many broadleaf (dicotyledonous) plants, where leaves fold up or droop at night and reopen during the day. Changes in turgor pressure mainly drive this movement, which is caused by the water pressure inside special cells called pulvinus cells located at the base of the leaf or leaflet. When these cells gain or lose water, they cause the leaf to move. These rhythmic changes are a response to environmental cues like light and temperature and serve several vital purposes [4, 19]. At night, folding the leaves helps reduce water loss through transpiration, which is especially beneficial in dry or tropical environments. It also protects the plant from cold or frost by minimizing exposure to harsh conditions. During the day, leaves reopen to capture more sunlight, which boosts photosynthesis, leading to better growth and energy production. Overall, this daily rhythm is a brilliant survival strategy that helps plants maintain balance (homeostasis) and perform at their best under changing environmental conditions [1, 11].

The rhythmic opening and closing of stomata, tiny pores on the surface of leaves, are essential for gas exchange between plants and the atmosphere. This process allows plants to take in carbon dioxide (CO_2) for photosynthesis and release oxygen and water vapor during transpiration. The term "syncopation of genomic oscillatory rhythms" is not scientifically accurate in this context; instead, this rhythmic behavior is best described as stomatal oscillation, which is regulated by the plant's circadian clock and environmental signals.

Stomata are surrounded by two specialized cells called guard cells, which control their movement. When guard cells absorb water, they swell and open the pore; when they lose water, they shrink and close it. This response is triggered by factors such as light, humidity, CO_2 concentration, and the plant's water status. Typically, stomata open during the day to allow CO_2 uptake for photosynthesis and close at night to minimize water loss through transpiration.

While the circadian rhythm sets the internal timing of stomatal movement, the plant continuously adjusts this rhythm based on real-time environmental conditions such as drought, high light intensity, or low humidity to balance gas exchange and water conservation [3, 20]. For example, under drought conditions, plants close their stomata early to prevent dehydration, even during daylight hours.

This precise control is vital for plant survival, especially under stress. It ensures that plants maintain efficient photosynthesis while conserving water, thereby enhancing resilience to environmental challenges like heat, dryness, and salinity. In short, stomatal regulation is a key adaptive mechanism that enables plants to optimize growth and survive in changing environments.

Rhythmic changes in leaf and stomatal behavior play a crucial role in enhancing plant fitness and yield by allowing plants to adjust their physiology in response to rapid environmental fluctuations dynamically. Stomatal regulation is directly linked to water-use efficiency, a critical trait for plant survival under water-limited conditions. By precisely timing the opening and closing of stomata, plants optimize the balance between carbon dioxide uptake for photosynthesis and water loss through transpiration. This coordination enables maximum photosynthetic activity during favorable periods while minimizing dehydration, particularly under drought stress [9, 18].

In addition to stomatal control, rhythmic leaf movements such as folding or repositioning help mitigate stress from excessive light or high temperatures. By adjusting leaf orientation or closing leaflets, plants can shield their photosynthetic tissues from photoinhibition and heat damage, thereby maintaining metabolic function under adverse conditions.

These oscillatory behaviors are not passive responses but represent an active, strategic regulation of plant-environment interactions. They reflect a sophisticated adaptation mechanism that allows plants to anticipate and respond to daily and seasonal environmental changes, optimizing resource use and stress resilience. This dynamic control enhances overall plant performance and productivity in fluctuating habitats.

Understanding and harnessing the rhythm-based movements of plants such as the daily opening and closing of stomata and the rhythmic folding or repositioning of leaves is crucial for improving crop yield and agricultural sustainability. Enhanced stomatal control allows plants to minimize water loss during transpiration while maintaining efficient carbon dioxide uptake for photosynthesis. This improved water-use efficiency is particularly beneficial in arid and drought-prone environments, where water conservation directly impacts plant survival and productivity.

Furthermore, crops with more efficient and precisely timed leaf movements can better manage stress from fluctuating environmental conditions, such as high light intensity, heat, or cold. These rhythmic behaviors enable plants to optimize light capture during favorable periods and protect photosynthetic tissues during stressful ones. Given the increasing challenges posed by climate change, including unpredictable weather patterns and prolonged droughts, such dynamic responses offer a powerful adaptive advantage.

In regions where water scarcity and climatic variability threaten food security, crop performance could be significantly improved by breeding or genetically engineering new crop varieties with more responsive and robust leaf and stomatal rhythms. These crops would be better synchronized with daily and seasonal environmental cycles, leading to higher growth efficiency and yield stability under stress conditions [14, 25]. By optimizing these natural rhythms, agriculture can move toward more resilient and sustainable production systems.

Leaf movements, such as the rhythmic folding or drooping of leaves (nyctinasty), serve not only physiological functions but also act as a defensive strategy against herbivores. By folding their leaves at night or under stress, plants can reduce the visibility and accessibility of their tissues, making them less attractive or harder to consume for herbivorous insects. Similarly, the rhythmic closure of stomata, especially in response to environmental signals such as darkness, drought, or pathogen-associated molecular patterns, can limit the entry points for microbial pathogens. Many bacteria and fungi exploit open stomata as a route to invade plant tissues, and timely stomatal closure is a key component of the plant's innate immune response. Understanding these dynamic behaviors opens new avenues for crop improvement, including the development of pest-resistant varieties and enhanced disease resilience through the genetic or environmental programming of rhythmic leaf and stomatal movements [8, 23].

Beyond individual plant survival, rhythmic movements play a role in shaping ecological interactions at the community level. The timing of stomatal opening and closure influences the local microclimate by modulating transpirational cooling, humidity, and CO_2 concentration around the plant. These microclimatic changes can affect the growth and physiological behavior of neighboring plants. Furthermore, synchrony or asynchrony in leaf and stomatal rhythms across different plant species can influence patterns of light interception, water use, and gas exchange, thereby affecting competition for resources and promoting species coexistence. Such interactions highlight the complex and often overlooked role of plant movement in ecosystem-level processes [7, 22].

The genetic and physiological mechanisms underlying these rhythmic movements are fundamental to plant survival, adaptation, and productivity. These include circadian clock-regulated gene networks, ion transport systems in guard cells, and turgor-driven motor functions in pulvini. Research into these mechanisms using genomic, transcriptomic, and proteomic approaches is uncovering key regulatory pathways that allow plants to anticipate and respond to environmental fluctuations [2, 10]. By deciphering how plants dynamically regulate their physiology, sometimes described as "respiratory gymnastics" in a figurative sense, we can harness their natural adaptability to optimize growth under stress conditions such as drought.

Ultimately, understanding the molecular and ecological basis of plant rhythmicity is not only critical for advancing plant biology but also for guiding sustainable agricultural practices and ecosystem management in the context of climate change and increasing anthropogenic pressures. This knowledge enables the design of smarter cropping systems, resilient crop varieties, and innovative strategies for food security.

2 The Mechanisms Behind Stomatal Movements

Stomata are tiny pores found on the surfaces of leaves and stems that play a vital role in regulating gas exchange in plants. They allow the uptake of carbon dioxide (CO_2) for photosynthesis and the release of oxygen (O_2) as a byproduct of respiration. Each stoma is formed by a pair of specialized guard cells, which are typically bean-shaped and surround the pore. These guard cells control the opening and closing of the stomatal pore through changes in turgor pressure, driven by the movement of ions (such as potassium and chloride) and water in and out of the cells. This process involves complex ion transport mechanisms and osmotic adjustments that enable the guard cells to swell (opening the pore) or shrink (closing the pore) in response to environmental signals like light, humidity, and CO_2 levels.

This precise regulation is essential for maintaining plant homeostasis, particularly in balancing photosynthesis and water loss through transpiration. By adjusting the size of the stomatal aperture, plants can optimize CO_2 intake for energy production while minimizing excessive water loss. This balance is especially critical in arid or water-limited environments, where conserving water is essential for survival. The structural and functional design of stomata, along with their dynamic responsiveness, makes them a key adaptation for plant growth, survival, and productivity under fluctuating environmental conditions [12, 16].

Stomata function as dynamic gatekeepers that regulate gas exchange and water loss in plants. Each stoma is flanked by two specialized guard cells that control the opening and closing of the pore in response to environmental and physiological signals. The mechanism relies on changes in turgor pressure within the guard cells.

When guard cells take up water through endosmosis, they become turgid and swell, causing the stomatal pore to open. This water influx is driven by a decrease in the guard cell's water potential, which results from the active accumulation of solutes, primarily potassium ions (K^+), chloride ions (Cl^-), and the organic anion malate^{2-}. These ions are transported into the guard cells via specific ion channels and transporters in the plasma membrane and tonoplast. The influx of K^+ and Cl^-, along with the synthesis of malate to maintain charge balance, lowers the osmotic potential inside the guard cells, drawing water in from neighboring epidermal cells.

Conversely, when the plant needs to conserve water, such as under drought or high salinity, these ions are actively exported from the guard cells. The efflux of K^+, Cl^-, and malate increases the water potential within the guard cells, leading to water loss and a reduction in turgor pressure. As the guard cells become flaccid, the stomatal pore closes.

This ion-driven mechanism allows plants to finely balance CO_2 uptake for photosynthesis with water loss through transpiration, maintaining internal homeostasis under fluctuating environmental conditions [6, 24].

The circadian clock is a fundamental biological system that regulates a wide range of plant physiological processes, including the daily opening and closing of stomata, tiny pores on the surface of leaves that control gas exchange and water loss. Plants possess an internal circadian clock that synchronizes their physiological activities with the 24-h cycle of light and darkness. This internal rhythm allows them to anticipate daily environmental changes rather than merely reacting to them.

Stomatal movement is one of the most essential circadian-regulated processes. Stomata typically open during the day to allow the uptake of carbon dioxide (CO_2) for photosynthesis and close at night to minimize water loss through transpiration. This rhythmic behavior is not just a passive response to light and dark but is actively controlled by the circadian clock, even under constant environmental conditions. The clock regulates gene expression and metabolic pathways in guard cells, the specialized cells that surround each stoma, enabling precise control over pore opening and closure [5, 15].

The length of the photoperiod the duration of daylight also plays a key role in stomatal regulation. Studies have shown that longer days can enhance stomatal conductance and photosynthetic activity, while short or irregular photoperiods may disrupt the timing of stomatal opening, leading to suboptimal gas exchange. This coordination ensures that stomata are fully functional during peak light periods, maximizing photosynthesis while conserving water.

When the circadian rhythm is disrupted, such as under fluctuating or artificial light and temperature conditions, plants often show reduced photosynthetic efficiency and poor water-use efficiency. This misalignment impairs their ability to respond effectively to environmental changes, ultimately affecting growth and productivity. Therefore, the circadian regulation of stomata is essential for plant fitness, allowing plants to optimize resource use and adapt to dynamic environments.

Stomata are the primary gateways for gas exchange in leaves, allowing plants to take in carbon dioxide (CO_2) for photosynthesis and release oxygen and water vapor. Their opening and closing are tightly regulated by environmental signals, with light and CO_2 concentration being among the most important.

Blue light is the key driver of stomatal opening. It is detected by specialized photoreceptors (such as phototropins) in the guard cells that surround each stoma. When blue light is absorbed, it activates H^+-ATPase enzymes in the guard cell membranes. These enzymes pump protons (H^+) out of the guard cells, creating an electrochemical gradient across the cell membrane. This gradient drives the influx of potassium (K^+) and chloride (Cl^-) ions into the guard cells.

As the concentration of these ions increases inside the guard cells, their osmotic potential decreases, causing water to move in through osmosis. The resulting increase in turgor pressure makes the guard cells swell and bend, leading to the opening of the stomatal pore. This process, known as blue light-induced stomatal opening, is essential for maximizing CO_2 uptake during the day, thereby enhancing photosynthesis [20, 22].

Red light also influences stomatal behavior, though its effect is less direct. Red light is primarily used in photosynthesis within guard cell chloroplasts, generating ATP and sugars that support the energy demands of ion transport and turgor regulation. While red light alone can induce slight stomatal opening, it plays a supporting role by sustaining the metabolic activity needed to keep stomata open after initial blue light activation. In many plants, continuous exposure to red light helps maintain stomatal aperture throughout the day by providing a steady energy supply, ensuring prolonged gas exchange for photosynthesis.

Blue light initiates stomatal opening through a well-defined signaling pathway, while red light supports this process by cellular metabolism required for sustained opening. Together, they ensure efficient gas exchange during daylight hours, optimizing plant growth and water-use efficiency.

To understand how plants balance gas exchange and water conservation, consider one key aspect of stomatal regulation: the role of carbon dioxide (CO_2) in controlling stomatal behavior. Stomata adjust their opening in response to the concentration of CO_2 inside the leaf, not just the atmosphere. When internal CO_2 levels are low, such as during active photosynthesis, the stomata tend to open wider to allow more CO_2 to enter the leaf for photosynthesis. Conversely, when CO_2 accumulates inside the leaf, the stomata close or partially close, because the plant has sufficient CO_2 for photosynthesis and no longer needs to keep the pores open.

This mechanism helps the plant maximize photosynthetic efficiency while minimizing water loss through transpiration. When stomata are open, water vapor escapes as CO_2 enters, so that prolonged opening can lead to dehydration. By closing under high internal CO_2 conditions, the plant conserves water without sacrificing photosynthetic performance.

Specialized sensors in the guard cells detect changes in CO_2 levels, triggering a cascade of signaling events. These signals alter the movement of ions (like potassium and chloride) in and out of the guard cells, which changes their turgor pressure and, consequently, the size of the stomatal pore. This allows fine-tuned control over gas exchange in response to metabolic demand and environmental conditions [8, 11].

In short, CO_2-mediated stomatal regulation is a critical adaptation that enables plants to optimize growth and survive under varying environmental conditions (Table 1).

Stomatal opening and closing are controlled by changes in turgor pressure within the guard cells that surround each pore. This process is driven by the movement of ions in and out of the guard cells, which alters their osmotic potential and causes water to enter or leave through osmosis.

Key ions involved include potassium (K^+), chloride (Cl^-), and the organic anion malate^{2-}. These ions are transported across the guard cell membranes by specific ion channels and transporters, which are activated in response to environmental signals such as light, CO_2 levels, and humidity. For example, blue light triggers the activation of proton pumps (H^+-ATPases), which hyperpolarize the membrane and drive the influx of K^+ ions into the guard cells [8, 20]. This accumulation of K^+ lowers the osmotic potential, causing water to flow in, increasing turgor pressure, and leading to stomatal opening.

To maintain electrical and osmotic balance, Cl^- ions also enter the guard cells. Additionally, malate is synthesized inside the cells to counterbalance the positive charge of K^+, ensuring stable ion composition and sustained turgor pressure. This coordinated regulation of ion fluxes allows the plant to precisely control stomatal aperture in response to both internal physiological status and external environmental conditions.

These processes are governed by complex signaling pathways that integrate multiple cues such as light intensity, drought stress, and CO_2 concentration ensuring that stomatal movements optimize gas exchange for photosynthesis while minimizing water loss.

Research in stomatal biology relies on model organisms to understand how guard cells control stomatal opening and closing, a process central to plant gas exchange. Stomata are tiny pores on leaf surfaces that allow plants to take in carbon dioxide (CO_2) for photosynthesis while releasing oxygen and water vapor. A combination of circadian rhythms and environmental signals such as light, humidity, and CO_2 levels tightly regulates their activity. These cues trigger osmotic changes within the guard cells, driven by the movement of ions like potassium (K^+), chloride (Cl^-), and malate through specific ion channels and transporters. This regulates water flow into and out of the cells, causing them to swell (opening the pore) or shrink (closing it).

This dynamic regulation allows plants to balance two critical needs: maximizing gas exchange for photosynthesis and minimizing water loss through transpiration. This balance is essential for plant survival and productivity across diverse and changing environments [2, 8, 9].

Table 1 Mechanisms behind stomatal movements

Mechanism/process	Key components	Trigger/stimulus	Functional role	Molecular pathways	Experimental evidence	Adaptive significance
Ion transport in Guard cells	Potassium (K^+) channels, Chloride (Cl^-) channels, Malate, Proton pumps (H^+-ATPases), Aquaporins	Light, CO_2 concentration, ABA, circadian clock	Regulates turgor pressure for stomatal opening/closing	Blue light activates ATPases, leading to K^+ and Cl^- uptake, osmotic water influx through aquaporins	Patch-clamp studies, fluorescence-based ion imaging	Maintains water balance and gas exchange under varying conditions
Blue light signaling pathway	Phototropins, 14-3-3 proteins, H^+-ATPase	Blue light	Initiates stomatal opening	Phototropins activate H^+-ATPase via 14-3-3 proteins, creating an electrochemical gradient	Mutant analysis of phototropins, infrared gas exchange techniques	Optimizes photosynthesis by increasing CO_2 uptake during daylight
Red light signaling pathway	Phytochromes, G-proteins, Photosynthetic electron transport chain	Red light	Supports stomatal opening through photosynthetic activity	Red light stimulates photosynthesis, lowering CO_2 concentration in guard cells, triggering opening	Studies with phytochrome mutants, chlorophyll fluorescence assays	Enhances CO_2 assimilation in optimal light environments
ABA-mediated stomatal closure	Abscisic acid (ABA), PYR/PYL/RCAR receptors, SnRK2 kinases, SLAC1 anion channels	Drought, water stress	Close stomata to conserve water	ABA binds PYR/PYL/RCAR, activates SnRK2 kinases, and phosphorylates SLAC1, leading to anion efflux and stomatal closure	Stomatal bioassays under ABA treatments, mutant studies	Improves drought resistance by reducing transpiration

(continued)

Table 1 (continued)

Mechanism/process	Key components	Trigger/stimulus	Functional role	Molecular pathways	Experimental evidence	Adaptive significance
Circadian regulation of stomata	Clock genes (TOC1, CCA1, LHY), Guard cell-specific circadian oscillators	Diurnal cycle, light–dark transitions	Synchronizes stomatal rhythms with day-night cycle	Clock genes regulate ion transporters and H^+-ATPases in guard cells	Real-time PCR for clock gene expression, time-lapse stomatal imaging	Enhances efficiency of photosynthesis and water use
CO_2 concentration response	CO_2-sensitive receptors, HT1 kinase, β-carbonic anhydrase	Ambient CO_2 levels	Adjusts stomatal aperture based on CO_2 availability	High CO_2 activates HT1 kinase, modulating ion channels in guard cells	Gas exchange measurements, CO_2 enrichment experiments	Balances CO_2 uptake for photosynthesis with water conservation
Temperature influence on stomata	Heat shock proteins (HSPs), Membrane fluidity regulators	Temperature fluctuations	Modulates stomatal aperture under heat stress	Elevated temperature alters membrane properties, activates HSPs, and influences ion channel activity	Thermal imaging, guard cell electrophysiology	Protects plants from overheating and dehydration
Humidity response	Aquaporins, ABA, Ion channels	Relative humidity	Adjusts stomatal conductance to ambient moisture levels	Low humidity triggers ABA synthesis, altering ion flux in guard cells	Gas exchange and humidity-controlled chamber studies	Maintains transpiration rates suitable for water and nutrient transport

(continued)

Table 1 (continued)

Mechanism/process	Key components	Trigger/stimulus	Functional role	Molecular pathways	Experimental evidence	Adaptive significance
Auxin-mediated movements	Auxins (IAA), PIN proteins, Guard cell-specific auxin transporters	Light gradients, developmental cues	Fine-tune stomatal aperture in response to light	Auxin transport modulates ion channel activity, influencing turgor pressure	Auxin biosynthesis mutants, fluorescent auxin reporters	Aligns stomatal function with developmental and environmental needs
Malate and solute accumulation	Malate, Sucrose, Vacuolar transporters	Light, circadian rhythms	Provides osmotic balance for stomatal movement	Synthesis of malate and sucrose increases osmotic pressure, driving water influx	Metabolomic profiling, mutant studies on vacuolar transporters	Supports rapid adjustments in stomatal dynamics
Hormonal cross-talk in stomatal function	ABA, Cytokinins, Ethylene, Jasmonates	Combined environmental stresses	Integrates multiple stress signals to regulate stomata	Hormonal signaling pathways converge on ion channel regulation	Mutant analysis, hormone application essays	Enhances resilience to multifactorial stress environments
Phosphorylation cascades	Kinases (OST1/SnRK2), Protein phosphatases (ABI1/ABI2)	ABA, light, circadian signals	Regulates the activity of ion channels and transporters	ABA triggers SnRK2 activation, leading to phosphorylation of ion channels	Western blotting for phosphorylated proteins, mutant analysis	Facilitates precise control of stomatal aperture under stress

(continued)

Table 1 (continued)

Mechanism/process	Key components	Trigger/stimulus	Functional role	Molecular pathways	Experimental evidence	Adaptive significance
Aquaporin regulation	Plasma membrane intrinsic proteins (PIPs), Tonoplast intrinsic proteins (TIPs)	Light, humidity, ABA	Modulates water flux in guard cells	Aquaporins adjust water permeability based on osmotic gradients	Fluorescence imaging of aquaporins, transgenic studies	Ensures efficient water movement to support dynamic stomatal responses
Photoreceptor signaling	Phototropins, cryptochromes, Phytochromes	Light intensity and wavelength	Mediates light-driven stomatal movements	Photoreceptors activate downstream signaling cascades affecting guard cell turgor	Spectral light experiments, photoreceptor mutants	Optimizes stomatal function for varying light conditions

Source [7, 8, 16, 20, 25]

A deeper understanding of the molecular and physiological mechanisms behind stomatal regulation offers great potential for improving crop performance, especially under climate change conditions where water scarcity and rising atmospheric CO_2 are major concerns. By optimizing stomatal behavior, scientists can develop crops with better water-use efficiency, higher yields, and greater resilience, supporting more sustainable agricultural practices.

3 Hormonal Regulation of Stomatal Movements

Abscisic acid (ABA) is a central plant hormone that plays a critical role in regulating stomatal closure, especially under drought stress. When plants experience water scarcity, ABA is rapidly synthesized in the roots and leaves and transported to the guard cells surrounding the stomatal pores. This hormone acts as a key signal to trigger stomatal closure, thereby reducing water loss through transpiration and helping the plant maintain hydration during periods of water deficit [7, 12, 20].

The ABA signaling cascade in guard cells is a well-orchestrated process. ABA binds to specific receptors such as PYR/PYL/RCAR proteins, which in turn inhibit protein phosphatase 2C (PP2C) enzymes like ABI1 and ABI2. This inhibition activates SnRK2 kinases (e.g., OST1/SnRK2.6), which phosphorylate downstream targets, including ion channels. This leads to the activation of anion channels (such as SLAC1 and QUAC1), causing efflux of chloride (Cl^-) and malate from the guard cells. The resulting membrane depolarization activates outward-rectifying potassium (K^+) channels (GORK), promoting K^+ efflux. The loss of these ions reduces the osmotic potential in guard cells, leading to water efflux via osmosis, loss of turgor pressure, and ultimately stomatal closure.

In addition to rapid physiological responses, ABA also regulates gene expression to enhance drought tolerance. It induces the expression of stress-responsive genes such as RD29A, RAB18, and NCED3 (a key enzyme in ABA biosynthesis), which contribute to the synthesis of osmoprotectants like proline and late embryogenesis abundant (LEA) proteins. These compounds help stabilize cellular structures and maintain osmotic balance under stress [7, 16, 25].

However, ABA signaling is tightly regulated to prevent excessive stomatal closure, which could limit CO_2 uptake and impair photosynthesis. Therefore, the balance between ABA biosynthesis, degradation, and sensitivity is crucial for optimizing the trade-off between water conservation and carbon fixation.

Besides ABA, other plant hormones also influence stomatal dynamics. Auxins, primarily known for promoting cell elongation, can promote stomatal opening by activating plasma membrane H^+-ATPases in guard cells. This leads to membrane hyperpolarization, facilitating the influx of K^+ and Cl^- ions, water entry, and increased turgor pressure. However, auxins can also modulate ABA biosynthesis under stress, linking growth and stress responses [8, 12, 20].

Cytokinins promote cell division and are also involved in stomatal regulation. They enhance stomatal opening by stimulating K^+ uptake through activation of inward-rectifying K^+ channels (KAT1/KAT2). Cytokinins often act antagonistically to ABA, counteracting its closure effect and promoting gas exchange even under mild stress. This hormonal crosstalk fine-tunes stomatal behavior in response to internal and external cues [7, 9].

The circadian clock further integrates these hormonal signals to align stomatal movements with the day-night cycle. In species like *Arabidopsis thaliana* and wheat (*Triticum aestivum*), the clock regulates the rhythmic expression of genes involved in ABA sensitivity and ion channel activity, ensuring stomata open during the day and close at night [2, 23, 24]. For example, the core clock gene TOC1 modulates ABA signaling components, while PRR7 and PRR9 influence stomatal conductance. This temporal regulation allows plants to anticipate environmental changes and optimize both photosynthesis and water-use efficiency.

Stomatal movement is governed by a complex interplay of hormonal signaling, environmental cues, and circadian regulation, ensuring that plants can adapt to fluctuating conditions while maintaining growth and survival (Table 2).

Abscisic acid (ABA) signaling plays a central role in helping plants respond to water stress, and this process is closely regulated by the diurnal cycle. ABA levels fluctuate throughout the day in a rhythmic pattern, allowing plants to fine-tune stomatal behavior in response to changing environmental conditions.

During the day, when light is available and photosynthesis is active, plants need to open their stomata to take in carbon dioxide. At the same time, transpiration rates are high due to heat and light. To balance this, ABA concentrations are generally lower in the morning, permitting stomatal opening and gas exchange. However, as the day progresses and conditions become hotter and drier, ABA levels rise, especially under water-limited conditions. This increase triggers stomatal closure, reducing water loss and helping the plant conserve moisture during peak stress periods [7, 9, 12].

At night, when photosynthesis stops (because there is no light) and transpiration rates are naturally low, ABA levels typically decrease. This allows stomata to reopen in many plant species, preparing them for gas exchange at dawn. However, this nighttime reopening depends on the plant's water status under prolonged drought. ABA remains high, and stomata stay closed to prevent excessive water loss.

This daily rhythm in ABA signaling is controlled by a network of enzymes that regulate both ABA biosynthesis (e.g., NCED enzymes) and degradation, and it is synchronized with the plant's circadian clock. This ensures that stomatal responses are not only reactive but also anticipatory, aligning water conservation with daily environmental cycles.

Under prolonged drought, ABA signaling becomes stronger and more sustained, maintaining stomatal closure and significantly reducing water loss. This adaptive mechanism is crucial for plant survival during extended dry periods [2, 23, 24].

The diurnal regulation of ABA enables plants to dynamically balance the need for carbon dioxide uptake and water conservation, optimizing growth and resilience in fluctuating environments.

Understanding the molecular mechanisms underlying these hormonal interactions and their regulation of stomatal behavior is crucial for developing strategies to improve plant resilience to water stress, particularly in climate change and increasing

Table 2 Diverse hormonal regulation of stomatal movements

Hormone name	Source of hormone	Mechanism of action	Effect on stomatal aperture	Key molecular players	Environmental triggers	Physiological significance
Abscisic acid (ABA)	Synthesized in chloroplasts of mesophyll cells	ABA binds to receptors (PYR/PYL/RCAR family) → Activates SnRK2 kinases → Phosphorylation of SLAC1 and KAT1 ion channels → Efflux of anions and inhibition of K^+ influx → Turgor loss in guard cells	Stomatal closure	PYR/PYL/RCAR, SnRK2, SLAC1, OST1, ROS	Drought, high salinity, low humidity	Prevents water loss during stress, enhances drought tolerance
Auxin (indole-3-acetic acid, IAA)	Synthesized in shoot apices, young leaves	Activates plasma membrane H^+-ATPase → Hyperpolarization of guard cell membrane → Activation of K^+ influx channels → Increased turgor pressure	Stomatal opening	Auxin-binding protein 1 (ABP1), H^+-ATPase, K^+ channels	Light, low CO_2	Promotes photosynthesis by facilitating gas exchange
Cytokinins	Synthesized in roots, transported via xylem	Modulates ROS and NO levels in guard cells → Regulates ion channel activity → Promotes turgor changes	Stomatal opening	Type-A and Type-B cytokinin response regulators, ROS, NO	Light, nutrient availability	Enhances gas exchange and nutrient uptake
Ethylene	Produced in all tissues, especially during stress	Ethylene signaling via EIN2/EIN3 pathway → Alters ROS/NO levels → Inhibits H^+-ATPase activity → Depolarization of guard cell membrane	Stomatal closure	EIN2, EIN3, ROS, NO	High ozone, pathogen attack	Reduces pathogen entry by limiting stomatal opening
Jasmonic acid (JA)	Synthesized in chloroplasts, peroxisomes	JA-Ile conjugate interacts with COI1 receptor → Modulates ion channel activity via ROS production	Stomatal closure	COI1, JAZ proteins, ROS, NO	Herbivory, pathogen attack	Reduces water loss and pathogen entry

(continued)

Table 2 (continued)

Hormone name	Source of hormone	Mechanism of action	Effect on stomatal aperture	Key molecular players	Environmental triggers	Physiological significance
Salicylic acid (SA)	Synthesized in chloroplasts	Modulates ROS and NO production → Alters ion channel activity in guard cells	Stomatal closure	NPR1, ROS, NO, SOD	Pathogen infection, high temperature	Enhances defense against biotrophic pathogens
Gibberellins (GAs)	Synthesized in young tissues, developing seeds	GA signaling via DELLA proteins → Modulation of H^+-ATPase activity → Activation of K^+ channels	Stomatal opening	DELLA proteins, H^+-ATPase, K^+ channels	Light, developmental cues	Supports growth by enhancing CO_2 uptake

Source [2, 23, 24]

water scarcity. By manipulating these signaling pathways, it may be possible to enhance crop productivity and water-use efficiency, ensuring food security in regions affected by drought and water scarcity [8, 9, 12].

4 Leaf Movements and Their Rhythms

Leaf movements are essential physiological responses that allow plants to adapt to changing environmental conditions. These movements, such as nyctinasty, heliotropism, and photonasty, are triggered by both internal biological rhythms and external cues like light and temperature. They play a vital role in optimizing photosynthesis, conserving water, and protecting plant tissues from stress, thereby enhancing survival and productivity.

Nyctinasty, often called "sleep movements," refers to the rhythmic folding or drooping of leaves at night and their reopening during the day. This behavior is commonly seen in legumes (e.g., *Mimosa pudica*, *Phaseolus vulgaris*) and some trees. The primary functions of nyctinasty include reducing water loss (transpiration) at night and protecting the photosynthetic machinery from potential damage during early morning light exposure. Although photosynthesis stops at night due to the absence of light, the reopening of leaves before dawn allows plants to capture morning sunlight quickly. More importantly, closing leaves at night may prevent the buildup of reactive oxygen species that could damage chloroplasts when light returns. These movements are regulated by the circadian clock, which enables plants to anticipate daily light–dark cycles. Environmental factors such as light intensity, temperature, and humidity fine-tune the timing and extent of these movements [14, 16, 22]. This internal timing system allows plants to adjust their leaf posture in new environments, improving their ability to manage light and conserve energy.

Heliotropism, or solar tracking, is the directional movement of leaves or flowers to follow the sun's path across the sky. Plants like sunflowers (*Helianthus annuus*) and certain legumes use this strategy to maximize light capture throughout the day. This movement is driven by differential growth in the leaf petiole (the stalk connecting the leaf to the stem), controlled by the plant hormone auxin. When light hits one side of the petiole, auxin accumulates on the shaded side, causing those cells to elongate more than the illuminated side. This imbalance makes the leaf bend toward the light. The process is mediated by phototropins, which are blue-light receptors in plant cells that detect directional light and initiate the signaling cascade leading to auxin redistribution [5, 25]. By continuously adjusting their leaf angles, heliotropic plants maintain high photosynthetic rates, which is especially beneficial in shaded or fluctuating light environments [10].

Photonasty refers to non-directional leaf movements in response to changes in light intensity or quality, rather than direction. For example, some plants close their leaves during midday when sunlight is strongest to reduce transpiration and prevent photoinhibition, a condition where excessive light damages the photosynthetic apparatus. Conversely, leaves reopen when light levels drop in the late afternoon to capture available light. These responses are common in plants growing under variable light conditions, such as those in forest understories [3, 9]. Photonastic movements are regulated by changes in turgor pressure within specialized motor cells at the base of the leaf, known as pulvini. The movement of ions like K^+ and Cl^- into and out of these cells alters osmotic balance, causing water to flow in or out and resulting in swelling or shrinking that drives leaf motion. While guard cells control stomatal movement, pulvinar cells are primarily responsible for photonastic leaf movements [16, 23].

Nyctinasty, heliotropism, and photonasty are distinct but complementary strategies that help plants optimize light use, conserve water, and avoid stress. These movements are governed by specific receptor systems, such as phototropins for light direction and circadian-regulated pathways for timing, and involve hormonal (auxin) and ion-driven mechanisms (Table 3). Together, they enable plants to respond to their environment dynamically, ensuring efficient growth and resilience under changing conditions [8, 10].

Nyctinasty, heliotropism, and photonasty are unique leaf action strategies for plants to efficiently absorb light, decrease transpiration, and avoid adverse environmental effects. Circadian-controlled nyctinastic movements prevent harm following exposure to harmful UV or red wavelengths overnight; heliotropism ensures leaves are positioned to the sun during peak photon-receiving hours, and photonasty allows for immediate adjustments to transient shifts in incident light intensity. The regulation of these movements is complex, involving the interactions of different light receptors, plant hormones such as auxins, and numerous environmental cues. These movements of leaves help plants adapt and optimize their growth and photosynthesis in their habitat, subsequently increasing chances for survival. So, understanding these movements and their mechanistic basis is essential to advance agriculture techniques and increase crop yield in a fast-changing environment and climate variability [5, 15, 16, 24].

5 Molecular Mechanisms of Leaf and Stomatal Rhythms

Rhythmic movements of leaves and stomata are vital for plant survival, enabling them to optimize photosynthesis, conserve water, and adapt to changing environmental conditions. These movements are regulated by a complex network of internal and external signals, including circadian clock genes, photoreceptors, ion transporters, and plant hormones, all working together to ensure that plants respond at the right time and in the right way to daily environmental cycles.

At the heart of this regulation is the plant's circadian clock, an internal biological timer that synchronizes physiological processes with the 24-h day-night cycle. Core clock genes such as CCA1, LHY, TOC1, and GI form a feedback loop that controls the timing of key functions like stomatal opening and leaf movement. This allows plants to anticipate daily changes, opening stomata during the day for gas exchange

Table 3 Different types of leaf movement

Type of leaf movement	Mechanism	Rhythmic pattern	Environmental triggers	Key molecular and cellular components	Examples of plant species	Adaptive significance
Nyctinasty (night movements)	Turgor changes in pulvini lead to leaf folding at night	Circadian rhythm	Light (day/night cycle), temperature	Phytochrome, circadian clock genes, turgor pressure in guard cells	*Mimosa pudica, Oxalis species, Albizia julibrissin*	Conserves water by reducing transpiration during the night
Phototropism	Differential cell elongation in response to light direction, controlled by auxin redistribution	Non-rhythmic	Light intensity and direction	Auxins (IAA), photoreceptors (e.g., phototropin), PIN proteins (auxin transporters)	*Arabidopsis thaliana, Helianthus annuus*	Maximizes light capture for photosynthesis
Seismonasty (touch movements)	Rapid turgor loss or gain in pulvini due to mechanical stimulation (e.g., touch)	Non-rhythmic	Touch, vibration, wind	Calcium channels, actin filaments, turgor pressure changes	*Mimosa pudica, Desmodium gyrans*	Protection from herbivores?? movement to discourage further touch
Gravitropism	Differential growth rates in the upper and lower parts of the leaf or petiole in response to gravity	Non-rhythmic	Gravity	Statoliths, auxins, gibberellins, ethylene	*Arabidopsis thaliana, Zea mays*	Ensures leaves orient for optimal light capture, and root growth into soil
Thigmonasty (response to mechanical stress)	Movement due to reversible changes in turgor pressure or growth inhibition	Non-rhythmic	Mechanical stress (e.g., wind, touch)	Ion channels (K^+, Ca^{2+}), actin filaments, turgor pressure	*Dionaea muscipula, Mimosa pudica*	Avoids damage from herbivores or physical stress

(continued)

Table 3 (continued)

Type of leaf movement	Mechanism	Rhythmic pattern	Environmental triggers	Key molecular and cellular components	Examples of plant species	Adaptive significance
Hydronasty (water movement)	Leaf movement in response to water availability or changes in humidity, often involving cell expansion	Circadian rhythm or triggered by environmental moisture	Water stress, humidity	Abscisic acid (ABA), reactive oxygen species (ROS), aquaporins	Water lilies (*Nymphaea*), some aquatic plants	Helps regulate water balance and gas exchange
Solar tracking (heliotropism)	Leaves or flowers move to track the sun, ensuring maximal light exposure	Circadian rhythm	Light intensity and position	Photoreceptors (phytochromes), auxins, actin filaments	Sunflower (Helianthus annuus), morning glory (Ipomoea)	Maximizes photosynthesis by orienting leaves toward sunlight
Circadian leaf movements	Rhythmic leaf folding or unfolding driven by the plant's internal clock	Circadian rhythm	Light/dark cycles, temperature	Circadian clock genes (e.g., TOC1, LHY), phytochrome	Mimosa pudica, Albizia julibrissin	Synchronizes leaf movements with day-night cycles to conserve energy and resources

Source [5, 15, 16, 24]

and closing them at night to reduce water loss rather than simply reacting to them [2, 8, 18].

Photoreceptors act as the plant's sensors for light, linking the internal clock to the external environment. Phytochromes detect red and far-red light and help align the circadian rhythm with the natural light–dark cycle. Cryptochromes respond to blue and UV-A light and play key roles in regulating both stomatal conductance and leaf movement, ensuring optimal light capture and stress avoidance. Another group of blue-light receptors, phototropins, control directional responses like heliotropism, where leaves reposition to follow the Sun. This coordination ensures efficient photosynthesis and thermal regulation throughout the day [14, 22].

The actual opening and closing of stomata depend on changes in turgor pressure within the guard cells. This is driven by the movement of ions such as potassium (K^+), chloride (Cl^-), and malate. When ions flow into the guard cells, water follows by osmosis, increasing turgor pressure and opening the pore. When ions leave, the cells lose water, and the stomata close. Key ion transporters like KAT1 (for K^+ influx) and SLAC1 (for anion efflux) are regulated by both the circadian clock and environmental signals such as light, CO_2, and humidity, allowing precise control over gas exchange and water use [16, 20].

Plant hormones also play a crucial role in fine-tuning these rhythms. Abscisic acid (ABA) is a primary stress hormone that triggers stomatal closure during drought, and the circadian clock regulates its production and sensitivity to optimize water conservation. Auxins influence leaf positioning by promoting cell elongation on one side of the leaf petiole, causing it to bend toward light (a process known as phototropism). Cytokinins and other hormones interact with the clock and ABA pathways to modulate stomatal behavior, ensuring a balanced response to multiple environmental cues [6, 7].

Together, these genetic, biochemical, and hormonal pathways form a highly integrated system that allows plants to maximize carbon gain, minimize water loss, and adapt to stress. Disruption of any part of this network such as mutations in clock genes or ion channels can impair stomatal function and reduce plant fitness and crop yield [10, 20].

Understanding these mechanisms provides a foundation for developing crops with improved drought tolerance, water-use efficiency, and productivity. By engineering or breeding plants with optimized circadian regulation and stomatal dynamics, scientists can create more resilient agricultural systems in the face of climate change (Table 4).

Table 4 Molecular mechanisms of leaf and stomatal rhythms

Rhythm type	Molecular mechanism	Key proteins and genes involved	Regulatory pathways	Role of circadian clock	Environmental inputs	Impact on leaf and stomatal function
Circadian leaf movements	Leaf folding and unfolding driven by the internal circadian clock, modulated by light and temperature cycles	Phytochrome (PHYA, PHYB), Timing of CAB 1 (TOC1), Late Elongated Hypocotyl (LHY), Circadian Clock Associated1 (CCA1)	Circadian clock genes (TOC1, LHY, CCA1), photoreceptors (phytochromes)	Circadian clock genes control leaf movement in response to light/ dark cycles, ensuring synchronization with day/night patterns	Light/dark cycle, temperature	Maximizes energy capture by adjusting leaf orientation to sunlight, and reduces water loss at night
Stomatal rhythms	Stomatal opening and closing controlled by internal circadian regulation, influenced by CO_2, light, and ABA	Slow Anion Channel 1 (SLAC1), Potassium Channel 1 (KAT1), Open Stomata 1 (OST1), ABA receptors (PYR1, PYLs)	ABA signaling, CO_2 sensing, circadian clock signaling	The circadian clock modulates stomatal behavior by influencing the expression of ion channels and enzymes that control guard cell turgor	Light, CO_2, humidity, temperature	Regulates gas exchange and water loss; optimizes photosynthesis during daylight hours
Nyctinasty (leaf movements)	Leaf movements are driven by turgor pressure changes in pulvini, controlled by circadian and light/dark cycles	Phytochrome receptors (PHY), circadian clock genes (LHY, CCA1), potassium channels	Phytochrome signaling, ABA, and calcium signaling pathways	The circadian clock regulates leaf movements by integrating environmental cues with the plant's internal timing mechanisms	Light, temperature	Reduces transpiration by closing leaves during the night, conserving water

(continued)

Table 4 (continued)

Rhythm type	Molecular mechanism	Key proteins and genes involved	Regulatory pathways	Role of circadian clock	Environmental inputs	Impact on leaf and stomatal function
Phototropism (leaf movement)	Leaf movement toward light, mediated by differential growth due to auxin redistribution	Auxin (IAA), phototropins (phot1, phot2), PIN1 (auxin efflux carriers)	Auxin signaling, photoreceptor pathways (phototropins)	Photoreceptors modulate auxin distribution, leading to differential elongation and phototropic response	Light intensity and direction	Ensures optimal exposure to sunlight for photosynthesis
Seismonasty (leaf movement)	Leaf folding in response to mechanical stimuli (e.g., touch, vibration), due to changes in turgor pressure	Calcium channels, actin filaments, ion channels (K^+, Ca^{2+}), ABA	Ion flux (Ca^{2+}, K^+), actin filaments, ABA signaling	Non-circadian regulation, rapid response to mechanical stimuli rather than rhythm	Mechanical touch, vibration, temperature changes	Protection against herbivores, rapid response to external stimuli
Gravitropism (leaf movement)	Leaf or petiole bending in response to gravity, regulated by differential growth rates	Auxins (IAA), Auxin Response Factors (ARF), PIN proteins, gibberellins, statoliths (amyloplasts)	Auxin signaling, gibberellin signaling, calcium-mediated signaling	Gravitropism is influenced by circadian regulation, as circadian rhythm modulates the sensitivity to gravitational cues	Gravity, light	Ensures proper orientation of leaves for light capture and root growth toward soil
Tropisms (leaf movement)	Growth responses of plant parts to directional stimuli, including phototropism and gravitropism	Auxins (IAA), cytokinins, brassinosteroids, PIN proteins, phototropins	Auxin redistribution, ethylene signaling, calcium signaling	Circadian clock modulates the sensitivity of the plant to directional stimuli like light and gravity	Light, gravity	Maximizes light exposure and optimizes plant growth orientation

(continued)

Table 4 (continued)

Rhythm type	Molecular mechanism	Key proteins and genes involved	Regulatory pathways	Role of circadian clock	Environmental inputs	Impact on leaf and stomatal function
Thigmonasty (Leaf Movement)	Leaf folding in response to touch or mechanical stress, mediated by ion channel activity	Calcium (Ca^{2+}) channels, potassium channels, actin filaments, ABA	Calcium ion influx, turgor pressure regulation	Non-rhythmic, rapid response to touch	Mechanical touch, temperature, wind	Deters herbivores, protects plant from physical damage
Solar Tracking (Heliotropism)	Leaf or flower movement to track sunlight, optimizing photosynthesis	Phytochromes, cryptochromes, phototropins, actin filaments	Photoreceptor signaling (phytochromes, phototropins), auxin redistribution	The circadian clock adjusts the sensitivity of photoreceptors to light, coordinating leaf movement with daily sunlight patterns	Light intensity, direction	Enhances photosynthesis by orienting leaves toward the sun
Stomatal Closure Rhythms	Stomatal closure during the night or water stress, regulated by ABA and the circadian clock	ABA, OST1, SLAC1, PYR1 receptors, ROS	ABA signaling, calcium influx, ROS production	Circadian clock regulates the expression of ABA receptors and ion channels to anticipate nighttime water loss	Light/dark cycle, drought, CO_2 levels	Reduces water loss during the night or under stress conditions

Source [2, 8]

6 Environmental Factors Influencing Rhythmic Movements

Light is known to be one of the most dominant external cues that drive rhythmic movements in plants, significantly mediating leaf and stomatal behavior. Light's properties, quality, intensity, and duration, collectively called the photoperiod, dictate physiological and molecular responses in a plant, calibrating its internal circadian clock with external environmental cycles. The light-activated signaling system of photoreceptors and their regulation of stomatal and leaf opening, namely the phytochromes, cryptochromes and phototropins, ensures that all their movements are in accord with resource capture and conservation. An illustration of this is that phytochromes, receptors for red and far-red (FR) light, control the opening and closing of the stomata. In contrast, blue light is perceived by cryptochromes and phototropins, stimulating proton pumps in guard cells to promote the stomatal opening. The magnitude of these responses is also controlled by light intensity; bright (high-intensity) light promotes maximal stomatal opening to match elevated demand for photosynthesis, whereas dim (low-intensity) conditions inhibit stomatal behavior from conserving water. Photoperiod also influences brief leaf movements such as nyctinasty, which leaves open at night to maximize light capture and fold during the day to prevent water loss [2, 3, 8, 24].

The other prominent environmental variable that strongly impacts circadian-adjusted activity is temperature. Temperature changes also coordinate signals that phase and amplitude the circadian clock, controlling leaf positioning and stomatal behavior. In normal circumstances, warmer temperatures increase daytime stomatal opening to enable gas exchange and photosynthesis. Conversely, the stomatal opening is typically restricted by low temperatures that decrease transpiration rates during periods of nutrient decline. The temperature effect on rhythm is mediated by integrating thermosensitive genes in the circadian regulatory network. For instance, the genes Timing of CAB Expression 1 (TOC1) and Circadian Clock-Associated 1 (CCA1) have been identified to display differing expression levels, depending on the temperature, contributing to the synchronization of the plant with its environment [9, 18].

Leaf movements (heliotropism) are also sensitive to temperature. When temperature increases, buds may angulate themselves to minimize needless absorption of radiation energy/heat, thus preventing thermal stress. Temperature-dependent circadian control of rhythmic movements is essential for maintaining physiological stability, particularly in conditions where daily temperature (and seasonal temperature) shifts are significant.

Humidity and water availability represent other modulators of stomatal dynamics that provide complex fittings for plant water-use efficiency. The stomatal opening is usually enhanced during high humidity due to reduced evaporative demand, leading to better gas partitioning and photosynthetic efficiency. In contrast, stomata can shut when humidity is low, which prevents cells from losing more water from transpiration than they absorb from the soil. The relationship between humidity and stomatal behavior is determined through guard cell turgor pressure changes and active influx and efflux of potassium (K^+) and chloride (Cl^-) ions. This regulation is additionally regulated via hormone signaling pathways, specific to that of abscisic acid (ABA), depending on water availability. During drought, ABA accumulates and overrides the circadian control of stomatal rhythms to promote water-use efficiency. It is tightly regulated in dry environments to preserve water for plant survival, and this hormonal response is critical for plant survival. Leaf movements, such as nyctinasty, may also respond to water availability, with drought leading to delay or suppression of movements to conserve water via transpiration. The influence of humidity, water availability, and stomatal responses illustrates how plants can adjust their physiology in response to environmental variability[7, 8, 10, 22, 23].

These daily changes in environmental factors, including light, temperature, humidity and water availability, have helped introduce the rhythmic behavior of leaves and stomata (tiny holes on the surface of leaves) across different plant lineages. These factors interact in complex ways, creating unique adaptive strategies in the plant and promoting its growth and survival. Some may be even more sensitive, especially those growing under desiccated regimes whose stomatal conductance may also decline in response to increased aridity. However, plants adapted to tropical conditions might favor light quality and intensity in their responses due to the high solar radiation levels received at these latitudes. Genetic diversity across species leads to differences in the composition and regulation of circadian clock genes, photoreceptors and ion transporters, fueling species-specific patterns of rhythmic behaviors. Apart from the environmental regulation, the plasticity of leaf and stomatal rhythms also allows plants to respond adaptively to short-term dynamics, e.g., rapid changes in solar radiation due to the coverage of clouds or temperature shifts during day-night transitions. This plasticity is essential to optimize metabolic efficiency and ensure the flourishing of plants across diverse ecological niches.

Integrating environmental cues to control rhythmic movements is crucial to plant adaptability, allowing for the maximization of resource acquisition and the minimization of stress. These internal circadian rhythms are entrained to environmental light and temperature cues such that leaf and stomatal movements are anticipatory rather than strictly reactive, allowing plants to precondition themselves in advance to predictable ecological stressors. A metabolic process, for example, stomatal opening, starts before dawn in many species, under the control of the circadian clock, in anticipation of the subsequent increase in light level. How is photosynthetic productivity adjusted with pre-coordinating by the maximum flux of CO_2 at the best condition of the environment early in the morning? Meanwhile, leaf movements such as heliotropism and nyctinasty are finely tuned to light and temperature cycles, optimizing the capture of plant resources while minimizing thermal and desiccation stresses [4, 13, 19, 21].

As the dominant environmental signal, light entrains rhythmic leaf and stomatal movements in a photoperiod, intensity and quality-dependent manner that must balance the physiologic consequences of a plant's behavior. Temperature serves as a Davos to circadian-driven migrations, tempering them to fit suitable daily and seasonal cycles. Environmental conditions, including humidity and water availability, regulate stomatal dynamics and are integrated with hormonal signaling pathways to improve water-use efficiency. Plants are very successful organisms that occupy diverse environments, and the plasticity of their rhythmic behaviors to dynamic environmental scenarios reflects this evolutionary success. The cross-talk between exogenous signals and internal regulatory processes highlights the importance of knowing these pathways to improve agricultural stress tolerance and yield under varying environmental scenarios (Table 5).

7 Ecological and Evolutionary Perspectives on Rhythmic Movements

Rhythmic movements of stomata and leaves are key adaptive strategies that help plants optimize resource use and thrive in diverse environments. These rhythms are controlled by the plant's internal circadian clock, which synchronizes physiological processes with predictable daily changes in light, temperature, and humidity. This allows plants to anticipate environmental shifts rather than react to them.

Stomatal behavior is a significant factor in water-use efficiency (WUE). In arid environments, plants regulate stomatal opening to balance the need for carbon dioxide (CO_2) uptake during photosynthesis with the need to minimize water loss through transpiration. Many plants open their stomata early in the morning when conditions are cool and humid, maximizing CO_2 intake while reducing water loss. In contrast, plants in tropical regions where water is abundant and light is high throughout the day often maintain high stomatal conductance for more extended periods, supporting sustained photosynthesis.

These habitat-specific adaptations also extend to leaf movements. For example, nyctinasty the rhythmic folding of leaves at night helps reduce water loss and protect against nocturnal herbivores, improving survival in dry or predator-rich environments [2, 5, 6, 15]. Similarly, heliotropism, where leaves track the sun's path, allows plants to maximize light capture, especially in shaded or competitive environments.

Over evolutionary time, these rhythmic behaviors have been shaped by selective pressures to balance carbon gain and water conservation. For instance, desert plants like succulents and Crassulacean Acid Metabolism (CAM) species open their stomata at night to fix CO_2 and close them during the day to prevent water loss, a highly effective adaptation for extreme dryness. In contrast, tropical plants have evolved to prioritize photosynthetic efficiency, with wide stomatal openings and pronounced heliotropic movements that exploit high light availability [9, 14, 18, 20, 23].

Stomatal rhythms are also closely linked to drought response. The hormone abscisic acid (ABA) plays a central role in closing stomata under water stress, and its interaction with the circadian clock ensures this response is timed for maximum efficiency. Leaf movements like nyctinasty and heliotropism further enhance water conservation by reducing leaf surface exposure during hot or dry periods [2, 10, 12, 16].

Table 5 Different environmental factors influencing rhythmic movements

Rhythmic movement type	Environmental factor	Effect on rhythmic movement	Molecular mechanism	Key proteins and genes involved	Regulatory pathways	Rhythmic pattern	Role of circadian clock	Physiological and ecological impact
Leaf movements (e.g., heliotropism)	Light (intensity/ direction)	Synchronizes leaf movements to maximize light absorption. Leaves orient toward light (solar tracking)	Photoreceptor-mediated response: phototropin activation leads to asymmetric cell growth	Phototropins, auxins, photoreceptor proteins (PHOT1, PHOT2), PIN proteins (auxin transporters)	Phototropism signaling; auxin redistribution	Diurnal, Circadian	Photoreceptors work with circadian clock to time leaf orientation for optimal photosynthesis	Enhances photosynthetic efficiency by optimizing leaf orientation for light capture
Stomatal opening/ closing	Light (intensity)	Light triggers stomatal opening to allow gas exchange for photosynthesis	Light induces proton pump activation, leading to an increase in the stomatal guard cell turgor	H^+-ATPases, photoreceptors (cryptochromes), K^+ channels	Light perception via photoreceptors leads to ion transport regulation	Diurnal, Circadian	The circadian clock aligns stomatal movements with light availability for efficient CO_2 uptake	Regulates gas exchange, optimizing photosynthesis and transpiration rates
Nyctinasty (leaf movement)	Dark/light cycle (day/ night)	Leaf movements synchronize with the light–dark cycle (e.g., closing at night)	Turgor changes in guard cells due to hormone signaling (e.g., ABA) and mechanosensitive proteins	Abscisic acid (ABA), ion channels (K^+, Ca^{2+}), PIP aquaporins	ABA signaling, turgor regulation pathways	Circadian, Diurnal	The circadian clock modulates sensitivity to light and darkness, ensuring timely leaf closing	It helps conserve water by closing stomata during the night, reducing water loss

(continued)

Table 5 (continued)

Rhythmic movement type	Environmental factor	Effect on rhythmic movement	Molecular mechanism	Key proteins and genes involved	Regulatory pathways	Rhythmic pattern	Role of circadian clock	Physiological and ecological impact
Stomatal movements (water stress)	Water availability (drought)	Decreases stomatal opening to prevent water loss	ABA accumulation in response to dehydration triggers guard cell closure	ABA receptors (PYR/PYL), S-type anion channels, SnRK2 protein kinases	ABA signaling pathway; ion flux control	Diurnal, Circadian	Circadian rhythms control ABA synthesis and signaling, optimizing stomatal response to water stress	Contributes to drought tolerance by minimizing water loss through controlled stomatal closure
Heliotropic leaf movements	Temperature (day/night)	Temperature variation triggers leaf movement to optimize heat exchange and photosynthesis	Thermal signaling pathways and thermosensory proteins activate differential growth	Thermosensitive proteins, auxin transporters	Auxin-mediated signaling; thermoresponsive pathways	Diurnal, Seasonal	Circadian rhythms coordinate with temperature fluctuations to control leaf movement	Enhances photosynthesis and optimizes leaf cooling under high temperatures
Seismonasty (touch response)	Mechanical stimuli (touch)	Rapid leaf movements in response to touch or vibration (e.g., *Mimosa* leaf folding)	Mechanosensory pathways lead to ion fluxes and turgor loss in leaf cells	Ion channels (Ca^{2+}, K^{+}), mechanosensory proteins	Calcium signaling, mechanotransduction pathways	Non-Rhythmic, Triggered by Stimulus	The circadian clock does not directly regulate it, but may have indirect modulation by hormonal pathways	Protective mechanism against herbivory or environmental stress

(continued)

Table 5 (continued)

Rhythmic movement type	Environmental factor	Effect on rhythmic movement	Molecular mechanism	Key proteins and genes involved	Regulatory pathways	Rhythmic pattern	Role of circadian clock	Physiological and ecological impact
Gravitropism (root/stem growth)	Gravity	Roots and stems reorient growth according to gravity	Gravitropic signaling involves auxin redistribution in response to gravity	Auxins, PIN proteins, gravity-sensing proteins	Auxin transport and redistribution	Diurnal, Circadian	Circadian regulation may optimize growth responses to gravity over time	Ensures correct root growth direction and stem orientation for optimal plant stability
Floral movements (e.g., opening/ closing)	Light (photoperiod)	Flowers open and close depending on light conditions to maximize reproductive success	Photoreceptor-mediated activation of genes regulating flower opening	CONSTANS, FT (Flowering Time) gene, photoreceptors	Photoreceptor signaling pathways CONSTANS signaling	Diurnal, Seasonal	The circadian clock and photoperiod interact to ensure synchronized flowering time	Increases reproductive success by aligning flower opening with optimal pollinator activity
Stomatal movements (CO_2 regulation)	CO_2 concentration	Low CO_2 triggers stomatal opening; high CO_2 triggers closing	CO_2 influx into guard cells leads to changes in turgor via ion flux regulation	K^+ channels, ABA, guard cell enzymes	CO_2 detection through carbonic anhydrase, signaling pathways	Diurnal, Circadian	Circadian rhythm modulates CO_2 sensing and stomatal movement throughout the day	Optimizes CO_2 uptake for photosynthesis while minimizing water loss

Source [2, 5, 6, 15]

These rhythmic adaptations have broader ecological and agricultural significance. In arid ecosystems, plants with efficient water conservation strategies help prevent soil erosion and support biodiversity. In tropical forests, high photosynthetic rates driven by rhythmic stomatal and leaf behaviors sustain high primary productivity and complex food webs.

For agriculture, understanding these natural rhythms opens the door to developing crops with improved drought tolerance, higher WUE, and greater yield stability under climate change. By breeding or engineering plants with optimized circadian regulation, we can create more resilient crops tailored to specific environmental conditions.

The evolution of rhythmic stomatal and leaf movements reflects a fundamental trade-off between water conservation and photosynthetic performance, finely tuned by the interplay of environmental cues and internal regulatory networks. This deep understanding is crucial for advancing both ecological science and sustainable agriculture in a changing world [8, 14, 23].

8 Agricultural Implications of Leaf and Stomatal Rhythms

Stomatal and leaf movements are closely linked, and their underlying molecular controls are not independent. These rhythmic behaviors are governed by the plant's circadian clock, which allows plants to anticipate daily changes in light, temperature, and humidity, rather than just reacting to them. This internal timing system enables plants to fine-tune physiological processes for optimal performance.

For example, stomatal rhythms regulate gas exchange by opening during favorable conditions to allow carbon dioxide (CO_2) uptake for photosynthesis and closing to reduce water loss through transpiration. This balance is crucial for maintaining high water-use efficiency (WUE), especially in dry environments. Plants often open their stomata early in the morning when temperatures are cooler and humidity is higher, maximizing CO_2 intake while minimizing water loss. This synchronization with environmental conditions enhances photosynthetic efficiency, leading to greater biomass production and improved crop yields [5, 6, 15].

Understanding stomatal behavior is key to improving crop water management. By aligning irrigation with the natural rhythm of stomatal opening, farmers can deliver water when plants need it most, early in the day, just as stomata begin to open. This approach reduces evaporation and prevents over-irrigation, which can lead to problems like root hypoxia and nutrient leaching. Modern precision irrigation systems already use real-time environmental data and stomatal rhythm models to optimize water use, ensuring crops stay healthy while conserving resources—significant in water-scarce regions.

The future of agriculture lies in breeding crops with optimized rhythmic traits. While traditional breeding has focused on yield and stress resistance, new programs are now targeting circadian-regulated traits such as stomatal oscillation and leaf movement. For instance, crops engineered to close their stomata rapidly during midday heat can reduce water loss without shutting down photosynthesis completely. Similarly, enhancing heliotropism, the ability of leaves to track the sun, can improve light capture in dense plantings, boosting yield. Advances in genomics and gene editing tools like CRISPR-Cas9 allow scientists to precisely modify genes that control these rhythms, accelerating the development of climate-resilient varieties.

These principles also apply to controlled environments like greenhouses, where light, temperature, and humidity can be fine-tuned to match a plant's internal clock. By mimicking natural day-night cycles, growers can enhance growth, reduce stress, and increase productivity. Automated systems equipped with sensors and AI can adjust environmental conditions in real time, optimizing leaf positioning and stomatal function for maximum efficiency. In open fields, precision agriculture technologies such as drones, IoT sensors, and remote monitoring can integrate rhythmic plant behavior models to enable site-specific delivery of water and nutrients, improving resource use and reducing environmental impact [3, 18, 20].

However, climate change poses a significant challenge by disrupting the synchrony between plant rhythms and environmental cues. Rising temperatures, shifting rainfall patterns, and increased CO_2 levels can alter stomatal behavior and decouple it from optimal photosynthetic periods. While higher CO_2 may reduce the need for prolonged stomatal opening, potentially increasing WUE, it can also interfere with natural rhythms, requiring careful management to maintain productivity [2, 10, 22].

To address these challenges, a combination of genetic and agronomic strategies is needed. Breeding programs can select for heat-tolerant stomatal responses or robust circadian clocks that maintain rhythmicity under stress. CRISPR-based editing of clock genes like CCA1, LHY, or TOC1 offers a powerful way to engineer climate-resilient crops. At the same time, agronomic practices—such as adjusting planting times, using shade nets, or reflective mulches can help restore natural rhythms by reducing extreme light and heat exposure.

Studying how different species have evolved rhythmic adaptations provides valuable insights. Desert plants, for example, have evolved mechanisms like CAM photosynthesis to fix CO_2 at night, minimizing water loss. Tropical species maximize photosynthesis under high light. These natural strategies can inspire new crop traits for regions facing climate extremes [1, 4, 19, 21].

Ultimately, harnessing the power of rhythmic plant movements goes beyond individual crop improvement. It supports sustainable agriculture by enhancing resource efficiency, reducing environmental impact, and building resilience. As global pressures from climate change, population growth, and water scarcity intensify, understanding and applying the principles of plant rhythms will be essential for achieving long-term food security.

9 Technological Approaches to Studying Leaf and Stomatal Rhythms

Advances in molecular and imaging technologies have dramatically deepened our understanding of leaf and stomatal movements, going far beyond traditional large-scale observations. These tools now allow us to study the dynamic, real-time mechanisms behind rhythmic plant behaviors such as nyctinasty (nighttime leaf folding) and heliotropism (sun-tracking) in living plants, offering critical insights into how plants adapt to changing environmental conditions.

One of the most effective approaches for studying these movements is time-lapse photography. This technique captures images at regular intervals, enabling researchers to visualize and quantify the timing, amplitude, and direction of leaf motion over hours or days. By analyzing these sequences, scientists can extract data on movement patterns, angles, and velocities across populations of plants, providing a clear picture of how light, temperature, and circadian rhythms influence behavior [8, 9, 18, 20, 23].

Chlorophyll fluorescence imaging is another powerful method, beneficial for assessing photosynthetic activity and stomatal function. The ratio of variable to maximum fluorescence (Fv/Fm) reflects the efficiency of Photosystem II, and changes in this parameter across the leaf surface can reveal areas where stomatal conductance is reduced, often due to stress like drought or high light. This spatially resolved data helps identify early signs of stress and shows how daily stomatal rhythms impact overall plant productivity and resilience [7, 19].

For direct physiological measurements, infrared gas exchange systems are widely used. These instruments measure real-time CO_2 uptake, water vapor loss (transpiration), and stomatal conductance, linking stomatal behavior to photosynthetic performance. When combined with controlled environmental chambers, they allow researchers to test how factors like light intensity, humidity, and CO_2 concentration affect stomatal rhythms. Integrating this data with time-lapse or fluorescence imaging provides a comprehensive view of plant responses to environmental changes [11, 18].

At the molecular level, high-throughput genomics, including RNA sequencing (RNA-seq) and qPCR, has revealed the diurnal expression patterns of genes involved in stomatal and leaf movements. Key circadian clock genes such as Circadian Clock-Associated 1 (CCA1) and Timing of CAB Expression 1 (TOC1) regulate the timing of these processes by responding to environmental cues. RNA-seq studies have also identified rhythmic expression of ion channels, transporters, and enzymes in guard cells that control turgor pressure and stomatal aperture [17, 21].

These genomic insights are now being combined with CRISPR-Cas9 gene editing and chromatin immunoprecipitation (ChIP) assays to pinpoint how specific genes regulate rhythmic behaviors. ChIP helps map where transcription factors bind to DNA over the circadian cycle, revealing regulatory networks that drive stomatal opening and closure. CRISPR allows targeted modifications to these genes, enabling the development of crops with improved drought tolerance, water-use efficiency, and stress resilience [2, 10, 22].

Emerging real-time monitoring technologies are also transforming research and agricultural practice. Thermal imaging measures canopy temperature, which correlates with stomatal conductance. More superb canopies indicate open stomata and active transpiration, while warmer areas suggest water stress. This non-invasive method is ideal for large-scale field screening, helping to identify stress-tolerant crop varieties and optimize irrigation. When paired with machine learning algorithms, thermal data can uncover hidden patterns in plant responses to drought, heat, or salinity.

High-temporal-resolution sensors, including MEMS-based devices, enable continuous monitoring of stomatal conductance and leaf movement, integrating seamlessly with environmental sensors for real-time decision-making. These systems are key components of high-throughput phenotyping platforms, which automate the tracking of plant growth, leaf dynamics, and stress responses under controlled or field conditions. Such platforms are invaluable for screening crops with desirable rhythmic traits like enhanced water conservation or heat tolerance across diverse genotypes and environments.

Ultimately, these advances are not just for research; they are being applied in precision agriculture. Real-time stomatal data can optimize irrigation scheduling, ensuring water is used when stomata are closed (e.g., at night), reducing evaporation and improving efficiency. Similarly, fertilizer and pesticide applications can be timed to coincide with peak metabolic activity during the day, enhancing uptake and minimizing waste.

In summary, the integration of imaging, molecular biology, and sensor technologies has revolutionized our understanding of plant rhythms. These tools reveal the spatiotemporal dynamics of leaf and stomatal behavior and uncover their genetic control, offering powerful strategies to improve crop performance in the face of climate change, water scarcity, and increasing environmental stress.

10 Conclusion

Leaf and stomatal movements are rhythmically regulated processes that integrate plant physiology with the internal circadian clock and external environmental cues such as light, temperature, and humidity. These dynamic behaviors are orchestrated by a network of photoreceptors (including phytochromes, cryptochromes, and phototropins), hormonal signaling pathways, and ion transport mechanisms in guard cells. This coordination allows plants to optimize key physiological functions, particularly photosynthesis, gas exchange, and water-use efficiency (WUE), by aligning stomatal opening and leaf positioning with the most favorable times of day.

For example, stomata typically open during daylight to facilitate CO_2 uptake for photosynthesis and close at night to minimize transpirational water loss. This diurnal rhythm is not merely a passive response but is anticipatory, driven by the circadian clock, which enables plants to prepare for daily environmental transitions. Similarly, leaf movements such as nyctinasty (nighttime leaf folding) and heliotropism (sun-tracking) allow plants to maximize light capture while protecting photosynthetic machinery from excessive radiation or desiccation. These adaptive strategies are evolutionarily conserved across diverse plant species and ecological niches, reflecting their importance in balancing resource acquisition and stress avoidance.

Under increasing environmental pressures, especially drought and water scarcity, the regulation of stomatal rhythms becomes critical for crop survival and productivity. The plant hormone abscisic acid (ABA) plays a central role in this process, triggering stomatal closure under water stress by modulating ion channels in guard cells, thereby reducing turgor pressure and closing the pore. This mechanism helps conserve water without completely shutting down photosynthesis. Understanding these regulatory pathways opens opportunities for breeding or engineering crops with enhanced drought tolerance and improved water-use efficiency. For instance, modifying circadian clock genes such as Circadian Clock-Associated 1 (CCA1) and Timing of CAB Expression 1 (TOC1) using CRISPR-Cas9 technology could shift or extend stomatal opening to align with cooler, more humid periods, reducing water loss in arid environments.

Moreover, optimizing leaf and stomatal rhythms can significantly improve photosynthetic efficiency. In controlled environments like greenhouses, lighting and temperature can be fine-tuned to match a plant's natural circadian cycle, boosting biomass production. In field agriculture, integrating knowledge of plant rhythms into precision farming systems such as timed irrigation, nutrient delivery, and pest control can enhance resource use efficiency and reduce environmental impacts.

Climate change poses new challenges by disrupting the synchrony between internal clocks and external cues, such as altered temperature and photoperiod patterns. This desynchronization can impair stomatal function and reduce crop yields. However, advances in genomics, high-throughput phenotyping, and machine learning now allow researchers to identify genetic variants that confer resilience to these disruptions. By combining real-time monitoring (e.g., thermal imaging, chlorophyll fluorescence, and sensor networks) with predictive modelling, scientists can screen for crop genotypes with superior rhythmic traits under variable conditions.

Crucially, rhythmic regulation does not act in isolation. It is deeply interconnected with other hormonal networks such as those involving auxins, cytokinins, and gibberellins, which coordinate growth, development, and stress responses. Deciphering these interactions offers new targets for improving plant resilience to combined stresses, such as drought and heat, which are expected to increase with climate change.

At the ecosystem level, leaf and stomatal rhythms influence carbon and water fluxes, affecting regional and global biogeochemical cycles. Incorporating these dynamics into climate and ecosystem models improves predictions of vegetation responses to environmental change.

In summary, the rhythmic regulation of leaf and stomatal movements is a fundamental aspect of plant biology with profound implications for agricultural sustainability, food security, and climate resilience. By bridging fundamental discoveries in genetics and molecular biology with cutting-edge monitoring and breeding technologies, researchers can develop crops that use resources more efficiently, adapt to changing climates, and support sustainable food systems. The future of agriculture lies in harnessing the power of plant rhythms to create more intelligent, more resilient cropping systems in a resource-limited world.

References

1. Chang Y, Shi M, Sun Y, Cheng H, Ou X, Zhao Y, Zhang X, Day B, Miao C, Jiang K (2023) Light-induced stomatal opening in Arabidopsis is negatively regulated by chloroplast-originated OPDA signaling. Curr Biol 33:1071–1081.e5. https://doi.org/10.1016/j.cub.2023.02.012
2. Chaves MM, Costa JM, Zarrouk O, Pinheiro C, Lopes CM, Pereira JS (2016) Controlling stomatal aperture in semi-arid regions—the dilemma of saving water or being cool? Plant Sci 251:54–64. https://doi.org/10.1016/j.plantsci.2016.06.015
3. Chen C, Xiao Y-G, Li X, Ni M (2012) Light-regulated stomatal aperture in arabidopsis. Mol Plant 5:566–572. https://doi.org/10.1093/mp/sss039
4. Gago J, Daloso DM, Carriquí M, Nadal M, Morales M, Araújo WL, Nunes-Nesi A, Perera-Castro A V, Clemente-Moreno MJ, Flexas J (2020) The photosynthesis game is in the "inter-play": Mechanisms underlying CO_2 diffusion in leaves. Environ Exp Bot 178:104174. https://doi.org/10.1016/j.envexpbot.2020.104174
5. Horrer D, Flütsch S, Pazmino D, Matthews JSA, Thalmann M, Nigro A, Leonhardt N, Lawson T, Santelia D (2016) Blue light induces a distinct starch degradation pathway in guard cells for stomatal opening. Curr Biol 26:362–370. https://doi.org/10.1016/j.cub.2015.12.036
6. Hou S, Rodrigues O, Liu Z, Shan L, He P (2024) Small holes, big impact: Stomata in plant–pathogen–climate epic trifecta. Mol Plant 17:26–49. https://doi.org/10.1016/j.molp.2023.11.011
7. Kirkham MB (2023) Chapter 24 - Stomatal anatomy and stomatal resistance. In: Third E (ed) Kirkham MBBT-P of S and PWR. Academic Press, pp 473–496
8. Kong D, Hu H-C, Okuma E, Lee Y, Lee HS, Munemasa S, Cho D, Ju C, Pedoeim L, Rodriguez B, Wang J, Im W, Murata Y, Pei Z-M, Kwak JM (2016) L-Met activates arabidopsis GLR Ca^{2+} channels upstream of ROS production and regulates stomatal movement. Cell Rep 17:2553–2561. https://doi.org/10.1016/j.celrep.2016.11.015
9. Li J, Tian W, Chen T, Liu Q-Y, Wu H-W, Liu C-H, Fang Y-Y, Guo H-S, Zhao J-H (2024) N6-methyladenosine on the natural antisense transcript of NIA1 stabilizes its mRNA to boost NO biosynthesis and modulate stomatal movement. Mol Plant. https://doi.org/10.1016/j.molp.2024.12.011
10. Liu L, Zhou J, Zhang J, Zhong Y, Wang L (2024) MdBAM17, a novel member of the β-amylase gene family, positively regulates starch degradation in ALA-induced stomatal opening in apple (Malus × domestica). Hortic Plant J. https://doi.org/10.1016/j.hpj.2024.05.006
11. Maier-Maercker U (1984a) Turgor-mediated leaf movements in analogy with stomatal function and under the general aspect of water flux through the plant: I. Microautoradiographic localization of 86Rb and 43K in the laminar pulvinus of Phaseolus vulgaris L. J Plant Physiol 115:405–418. https://doi.org/10.1016/S0176-1617(84)80039-5
12. Maier-Maercker U (1984b) Turgor-mediated leaf movements in analogy with stomatal function and under the general aspect of water flux through the plant: II. Rhythmic transport of 86Rb and 43K in Trifolium repens L. and Oxalis acetosella L. J Plant Physiol 115:419–425. https://doi.org/10.1016/S0176-1617(84)80040-1
13. Muraoka Y, Yang G, Munemasa S, Takeuchi Y, Ishimaru Y, Murata Y, Uozumi N, Ueda M (2023) An outward-rectifying plant K+ channel SPORK2 exhibits temperature-sensitive ion-transport activity. Curr Biol 33:5488–5494.e7. https://doi.org/10.1016/j.cub.2023.10.057

14. Niu L, Wu X, Liu H, Hu X, Wang W (2024) Leaf starch degradation by β-amylase ZmBAM8 influences drought tolerance in maize. Carbohydr Polym 345:122555. https://doi.org/10.1016/j.carbpol.2024.122555
15. Oikawa T, Ishimaru Y, Munemasa S, Takeuchi Y, Washiyama K, Hamamoto S, Yoshikawa N, Mutara Y, Uozumi N, Ueda M (2018) Ion channels regulate nyctinastic leaf opening in Samanea saman. Curr Biol 28:2230–2238.e7. https://doi.org/10.1016/j.cub.2018.05.042
16. Rehman TU, Zhang L, Wang L, Ma D, Maki H, Sánchez-Gallego JA, Mickelbart M V, Jin J (2020) Automated leaf movement tracking in time-lapse imaging for plant phenotyping. Comput Electron Agric 175:105623. https://doi.org/10.1016/j.compag.2020.105623
17. Sun Z, Wang X, Song Y, Li Q, Song J, Cai J, Zhou Q, Zhong Y, Jin S, Jiang D (2023) StomataTracker: revealing circadian rhythms of wheat stomata with in-situ video and deep learning. Comput Electron Agric 212:108120. https://doi.org/10.1016/j.compag.2023.108120
18. Triozzi PM, Brunello L, Novi G, Ferri G, Cardarelli F, Loreti E, Perales M, Perata P (2024) Spatiotemporal oxygen dynamics in young leaves reveal cyclic hypoxia in plants. Mol Plant 17:377–394. https://doi.org/10.1016/j.molp.2024.01.006
19. Wang Z, Li L, Khan D, Chen Y, Pu X, Wang X, Guan M, Rengel Z, Chen Q (2023) Nitric oxide acts downstream of reactive oxygen species in phytomelatonin receptor 1 (PMTR1)-mediated stomatal closure in Arabidopsis. J Plant Physiol 282:153917. https://doi.org/10.1016/j.jplph.2023.153917
20. Xiao L, Ma W, Zhang J, Pu X, Rengel Z, Song Z, Chen Q (2023) Phytomelatonin interferes with flavonols biosynthesis to regulate ROS production and stomatal closure in tobacco. J Plant Physiol 284:153977. https://doi.org/10.1016/j.jplph.2023.153977
21. Yang S, Liang K, Sun Y, Zhang J, Cao Y, Zhou J, Wang A, Zhang L (2023) Evidence of the predominance of passive symplastic phloem loading and sugar transport with leaf ageing in Camellia oleifera. Hortic Plant J 9:811–825. https://doi.org/10.1016/j.hpj.2023.04.009
22. Yu H, Zhang J, Zhang S, Han Z (2023) Bionic structures and materials inspired by plant leaves: A comprehensive review for innovative problem-solving. Prog Mater Sci 139:101181. https://doi.org/10.1016/j.pmatsci.2023.101181
23. Yu T, Zhang J, Cao J, Cai Q, Li X, Sun Y, Li S, Li Y, Hu G, Cao S, Liu C, Wang G, Wang L, Duan Y (2021) Leaf transcriptomic response mediated by cold stress in two maize inbred lines with contrasting tolerance levels. Genomics 113:782–794. https://doi.org/10.1016/j.ygeno.2021.01.018
24. Zhang L, Zhao X, Wang Y-J, Zhang X (2009) Crosstalk of nitric oxide with Ca^{2+} in stomatal movement in Vicia faba guard cells. Acta Agron Sin 35:1491–1499. https://doi.org/10.1016/S1875-2780(08)60097-8
25. Zhao X, Wang Y-L, Wang Y-J, Wang X-L, Zhang X (2008) Extracellular Ca^{2+} regulating stomatal movement and plasma membrane K^+ channels in guard cells of Vicia faba under salt stress. Acta Agron Sin 34:1970–1976. https://doi.org/10.1016/S1875-2780(09)60016-X

Chapter 7
Light and Plant Life: Photoperiodism and Photomorphogenesis

Prasann Kumar, Amit Raj, Shipa Rani Dey, Joginder Singh, and Neha Gogia

1 Introduction

Light is one of the most critical environmental factors affecting the performance of plants, and also serves as a resource and signal. Plants, as members of the group of autotrophic organisms, mainly depend on light to acquire the energy necessary for synthesising organic compounds through photosynthesis [55]. Besides the synthesis of food, light controls growth and development processes such as germination of a seed, growth of a stem, growth of a leaf, flowering of a plant and the biological clock. Plant response to their environment depends on the intensity and wavelength of light, as well as its duration and direction. It is known that plants can perceive diverse light qualities, such as red light, far-red, blue, ultraviolet light, through specific molecules called photoreceptors [10]. Some of these are phytochromes involving red and far-red light, cryptochromes involving blue light, phototropins and UVR8 involving ultraviolet light, all regulating plant growth and development. This light quantity depends on geographical coverage, season and time of the day, thus, plants need to harness the changing light conditions [40]. Research has shown plants' ability to adapt to light to survive and grow optimally under different conditions. Appreciation

P. Kumar (✉) · A. Raj
Department of Agronomy, School of Agriculture, Lovely Professional University, Phagwara, Punjab, India
e-mail: prasann0659@gmail.com

S. R. Dey
Department of Agronomy, Faculty of Agricultural Sciences, GLA University, Mathura, Chaumuhan, Uttar Pradesh, India

J. Singh
Department of Botany, Nagaland University, Nagaland, India

N. Gogia
Fort Lee, NJ, USA

A. Kothari Chhajer and V. Singh Gour (eds.), *Introduction to Rhythms in Plants*, Studies in Rhythm Engineering, https://doi.org/10.1007/978-981-95-3592-7_7

of these responses impacts agriculture, horticulture and controlled environment plant production.

The fundamental feature of all plant life is the ability to harvest light and convert it into energy in photosynthesis. Chlorophyll, which is the most critical pigment within plants, absorbs light most efficiently in the blue and red ranges of the light spectrum, those two ranges being commonly referred to as the 400–500 nm and 600–700 nm regions, respectively, Chlorophyll absorbs blue and red light, whereas the light is not absorbed in the green region. This scattered green light is reflected by the inner structure of the leaves and escapes as reflected light; hence, the leaves have a green appearance to them despite having all the other spectral colours being highly absorbed. The absorbed light energy is used to develop ATP and NADPH, which would be used to synthesise the organic compounds needed for the plant's growth and development [32]. Photomorphogenesis can be defined as the process that is concerned with the development of plant shape and the formation of plant organs. Amphiarthros are plants with large hypocotyls primarily developed in a lightless environment, closed cotyledons and no chlorophyll, referred to as skotomorphogenesis [88]. As soon as it is placed under light conditions, it enters a process called de-etiolation, in which the plant undergoes photosynthetic maturity and organised leaves with chloroplasts and broader stems rather than elongated ones. Phototropism affects plants, which can determine the direction of the light and orient themselves in that direction. This process is known as phototropism, where phototropins are the primary receptors at the blue wavelength of light. Shoots of plants are positively phototrophic, and roots are negatively phototrophic because the former grow towards the light source. In contrast, the latter grows away from the source in search of water and nutrients in the soil [27]. Light–controlled biological rhythms are internal messengers that dictate the activities of various physiological processes in the plant due to daily and seasonal fluctuations. These diurnal rhythms affect factors such as stomata opening, photosynthesis and flowering processes to ensure the time and adjust appropriately to the environment. Photoperiodism is one of the well-documented factors for regulating flowering time and depends on light [53]. Some plants that require a long duration of darkness to flower are rice and chrysanthemum, whereas wheat and spinach bloom in short darkness periods. Tomatoes are categorised as day-neutral plants and are not affected by day length. This response is through the phytochrome system, which initially senses red/far-red light changes and converts the signal to affect flowering-related genes.

Photoperiodism refers to the physiological reaction of plants to the length of day and night. Addressing seasonal changes in flowering, variations in dormancy and other developments is vital. The photoperiodic response is, therefore, based on the perception of the number of hours of light through the phytochrome system. Photomorphogenesis refers to the growth and development processes that are initiated through light and or influenced through light and do not include the photosynthesis process [86]. They involve processes such as seed germination, stem growth, leaf size enlargement and flowering, among other processes. Photomorphogenesis, unlike photoperiodism, is chiefly linked to the structure and intensity of light signals. Light has been discovered to impact the plant life process and has been under research for many years since ancient times. The first documented experiments about the reactions of plants to light were conducted by Charles Darwin and his son, Francis Darwin, in the nineteenth century. Regarding phototropism, Grass seedlings were used as the sample; they discovered that the lipid sheath covering the shoot apex was the sensitive organ for perceiving light and produced signals for bending to begin. This process enabled them to reveal the existence of auxin, the plant hormone responsible for phototropic movement [83]. Julius von Sachs, the German botanist, significantly contributed to analysing the light utilisation in chlorophyll and its ability to carry out photosynthesis. He confirmed that chlorophyll is not present homogeneously in plant tissues but is concentrated in bodies called chloroplasts. He also proved that light is required for synthesis of chlorophyll [42]. Later, in the middle of the twentieth century, Sterling Borthwick and Harry Hendricks came across phytochrome, a photoreceptor of red and far-red light. They described how plants measure day length and the effects of photoperiodism on flowering, germination, etc. [30]. Due to discoveries in molecular biology and genetic engineering, the complexities of genetics and the biochemistry of light perception have been discovered. Cryptochromes and the more developed picture of phytochrome signalling have complemented the knowledge about how plants perceive light at the molecular level [24]. In addition to energy, light is an indispensable signal for plants to grow, develop and survive. Knowledge of how plants respond to light is relevant to several fields, ranging from farming to plant breeding, plant conservation and botany. Past and recent scientific studies are attributed to how these perceive light and respond to it to improve crop productivity and efficient agriculture. Scientific experiments conducted in the past and recent times have contributed a great deal in understanding the nature of perception and response of plants to light, but it has been established that the response of plants to light is quite intricate and uses photoreceptors like signalling mechanisms which control aspects of plant growth and development and metabolic productivity. Photoreceptors that mediate light perception by plants include phytochromes, cryptochromes, phototropins and UV Resistance Locus 8, which are sensitive to wavelengths. Signal transduction cascades that influence the germination of seeds, elongation of stems, expansion of leaves and development of chloroplasts as well as flowering time are triggered by these photoreceptors. Most recent studies have demonstrated that options to maximise photosynthetic potential and resource allocation in the field or in controlled environments could be provided by manipulating light quality, intensity and photoperiod to increase nitrogen assimilation, water-use efficiency and the ability to survive against stress in plants. This knowledge is finding application around controlled environment agriculture, where it is being used to exponentially improve practices such as light emitting diode (LED based spectral tuning that works towards co-ordinating plant light responses to crop management objectives. These types of knowledge become very important when it comes to precision agriculture since it allows specific interventions that lead to enhanced yields, quality and sustainability. In sum, the knowledge of the perceptual mechanisms and response to light is one of the foundations used to build future high-efficient agricultural systems.

2 Photoperiodism: Plants and Day Length Perception

Photoperiodism is one of the most basic and well-known phenomena that regulates the plant's development according to the amount of daylight. By operating light receptors and the circadian clock, they can measure photoperiod and significant events, including flowering, tuberization and dormancy. The CO-FT (CONSTANS–FLOWERING LOCUS T) regulatory module converts photoperiodic signals into developmental processes [77]. The molecular explanation is demonstrated by the controlled system of chrysanthemums grown under LED-modulating CEA or controlled environment agriculture [28]. Due to the recent articles indicating the basic mechanism of photoperiodism, future articles may have detailed information on improving the conditions favourable to plants that will increase crop production, challenging climates and markets. Manipulation of the environmental circumstances (duration of light, spectrum quality and temperature) to suit the needs of crop photoperiod. Breeding/engineer photoperiods adaptive varieties, which will succeed in difficult climates. By regulating flowering and harvest time to meet crop production needs in accordance to market demands.

2.1 Definition and Significance of Photoperiodism

2.1.1 Definition and Categories

Photoperiodism means the organism's response to the length of day or night, especially in plants. Photoperiodism is a vital plant response that helps the plant coordinate seasonal developmental processes. Plants may be grouped into short-day, long-day and day-neutral plants depending on their reaction to the length of a day [46]. In short-day plants (SDPs), flowering is induced when the night is longer than a specific period, while in long-day plants (LDPs), the opposite is the case. On the contrary, day-neutral plants flower at any photoperiod, they are instead regulated by developmental features or other factors such as temperature.

2.1.2 Ecological and Agricultural Significance

Photoperiodism regulates flowering, dormancy and tuberization to occur at the right time, which favours fertilisation and survival. For instance, in temperate environments, the timing of flowering is essential for synchronising with pollinators, hence seed release during the favourable weather conditions. In the framework of crops, knowledge of photoperiodism can be used to change the time of blooming and increase productivity [58]. This regulation is crucial for natural ecosystems and contemporary agriculture, where controlled environmental agriculture and selective breeding require photoperiod sensitivity.

2.2 *Molecular Basis of Photoperiodism*

2.2.1 Role of the Circadian Clock and Light Receptors

At the molecular level, photoperiodism, therefore, involves the differential regulation between the circadian clock and light receptors. An interaction between the internal timekeeping mechanism of a plant (circadian clock) and the capacity to detect light (light receptors) governs photoperiodism. They also determine the length and time of light by their internal clock. Such coordination enables them to sense day length and make important decisions such as when to start important developmental steps such as flowering. Circadian clock one of the known non-photosensitive timekeepers, this gives rise to oscillations in gene expression that persist under constant conditions and can be driven into synchronisation by light–dark cycles accordingly aligning internal plant physiology to the external environment. it connects the mechanism (oscillations + synchronisation) with the outcome (timing plant biology to the day-night cycle). [56]. Some examples of such light receptors include phytochromes and cryptochromes, which receive light signals of a given wavelength to regulate the circadian rhythm. These receptors allow plants to recognise the changes in the light quality and its duration and adjust the processes that occur during day and night periods.

2.2.2 Key Genes: CONSTANS (CO) and FLOWERING LOCUS T (FT)

Various genes are involved in photoperiodic regulation, and here, two genes stand out, and these are the CONSTANS (CO) gene and the FLOWERING LOCUS T (FT) gene. CO is a component of the circadian protein that regulates the circadian rhythm by regulating transcription in response to daylight. CO is synthesised and accumulates during short photoperiods, and its action leads to the transcription of a gene identified as FT, which codes for a mobile protein called the florigen [77]. FT then transports from the leaves to the shoot apical meristem, which promotes the shift from the vegetative phase to the reproductive phase. This CO-FT module is highly conserved in many plants and is a key point through which photoperiodic information is transduced to developmental programs (Fig. 1).

2.2.3 Interaction Between Light Receptors and Circadian Rhythms

The combination of light receptors and circadian rhythms accurately measures the day length. Light affecting phytochromes, which is red/far-red light, and cryptochromes for blue light, influence the stability and activity of CO. For instance, under long days, signals by light during the late afternoon synchronise with CO to promote the FT gene and flowering in long-day plants [72]. On the other hand, short-day plants do not allow the breakdown of CO during the specific evening period, thus

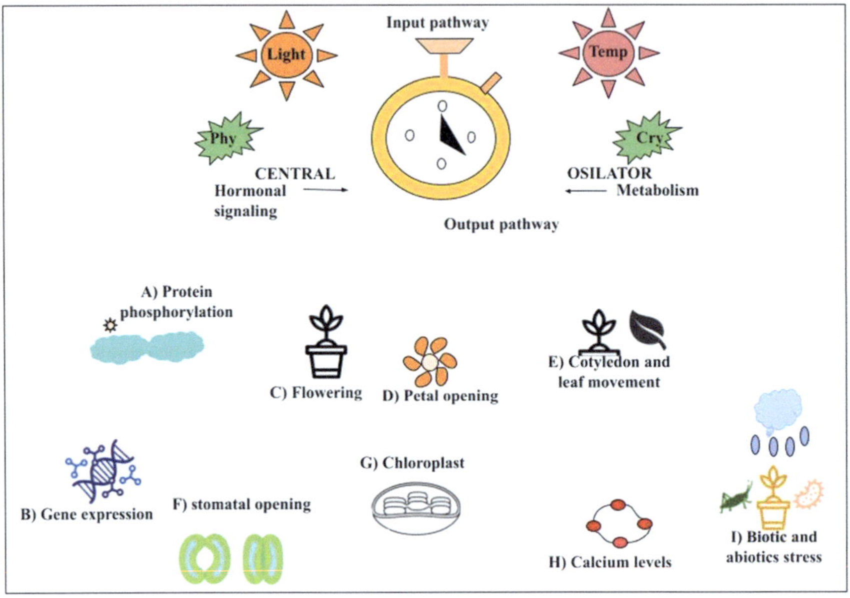

Fig. 1 GI (GIGANTEA) binds along with the CO promoter to activate the expression of *CO*

preventing the expression of FT during unfavourable conditions that prevent flowers from blooming. This precise mechanism can allow plants to determine the length of day and night, regulating most of their activities and processes.

2.2.4 Photoperiodism and Seasonal Adaptations

Besides flowering, photoperiodism also affects several other seasonal characteristics, such as dormancy and tuberization. Most perennials have short days in the autumn as days move on, thus preparing them for a dormant period since it is unfriendly for growth. Likewise, in crop plants such as potatoes, the light–dark period influences the tuberization to ensure tubers develop when circumstances are right for storing nutrients instead of growing more stems [73]. These adaptive responses emphasise the flexibility of photoperiodism as a molecular clock used to regulate the plant's development in response to seasonal changes.

2.2.5 Photoperiodic Response in Commercially Grown Chrysanthemums

In the recent study of commercially produced short-day chrysanthemum, authors clarified the method of manipulating photoperiod to strictly control and maximise the durations of flowering [93]. In this study, constant variation of photoperiod under the greenhouse environment was achieved by installing LED lighting systems. The length varies according to experimental design, but in most chrysanthemum photoperiod-control experiments under greenhouse conditions, LED lighting has been used several hours during the natural dark periods to extend the day length (2–6 h after sunset), or a night-break period (0.5–4 h in the middle of the night. By doing so, they achieved early flowering, which was forwarded to the crop cycle to respond to the market needs. From the molecular analysis, it was observed that during the elongation of the dark period, the amount of CO produced was less; therefore, a smaller amount of the FT gene was transcribed, resulting in the Flowering of the plant. [48]. It also helped to predict the yield and plan the flowering period according to market demands. With the combination of this technology and the more traditional breeding methods, one can start with the natural photoperiodic stimulation as the base of precision farming, which can bring both more efficiency and profitability of the same type of crop.

3 Phytochromes and Cryptochromes: Key Players in Photoperiodism

Phytochromes and cryptochromes are the two essential photoreceptors exclusively found in plants, controlling the developmental processes by regulating photoperiodic measurements. This red and far-red phytochromes and blue cryptochromes enable plants to perceive wavelengths at 660 nm, 730 nm and 320–500 nm, respectively, which make their alignment with the external environment [40]. The control of the localisation of the photoreceptors with the circadian clock enhances this process to maintain the proper timing of developmental signals. In this commentary, we expand on the details of phytochromes, how cryptochromes are involved in photoperiodism, and the molecular processes employed to convert light signals into developmental responses using a new study of rice crop enhancement [21].

3.1 Structure and Function of Phytochromes (Red/Far-Red Light Receptors)

Phytochromes are biliprotein photoreceptors that have two states: Pr form, which has red light-absorbing property with a red light at about 660 nm, and Pfr form with a far-red light at about 730 nm. Their structure consists of an apoprotein covalently bonded with a linear tetrapyrrole chromophore [57]. Upon absorption of red light, the chromophore part of the phytochrome changes its conformation and causes a transition from the inactive phytochrome state Pr to the active state Pfr. This photoactivation leads to other events that cause the copying and synthesis of materials required for various processes, including seed germination, stem extension and in this case,

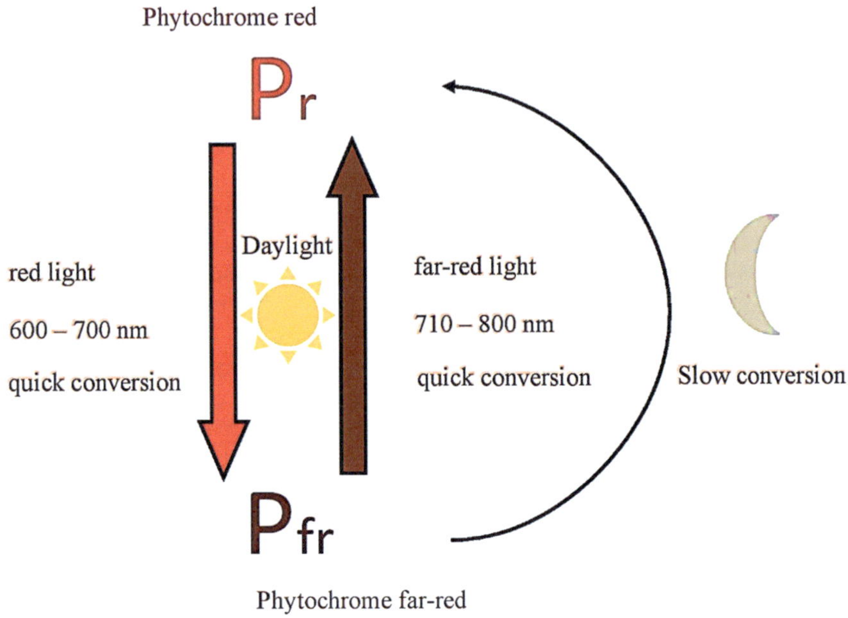

Fig. 2 Overview of phytochrome, its Pr and Pfr forms, light absorption

flowering. Pr/Pfr change allows plants to determine the light's quantity, quality and potential shading or changes in day length (Fig. 2).

3.2 Role of Cryptochromes (Blue Light Receptors) in Photoperiodic Responses

Cryptochromes are flavoproteins, which possess blue and ultraviolet-A (UV-A) light absorption properties, with a preferred absorption range of 320–500 nm; they use a flavoprotein, flavin adenine dinucleotide (FAD) as a chromophore. They are evolutionarily related to photolyases and together both form the photolyase/cryptochrome family (PHR); unlike photolyases, however, cryptochromes lack DNA-repair capabilities but are known to serve as photoreceptors to govern photomorphogenesis, circadian rhythms and photoperiodic flowering. PHR is a common domain for photolyases that bind to FAD, where the cofactor can undergo a conformational change after absorbing the light energy [24]. Despite the close relationship between cryptochromes and DNA photolyases in terms of structure, cryptochromes cannot repair DNA, instead, they are mainly involved in signalling. In making photoperiodic decisions, cryptochromes are involved in the changes in the expression of genes that control development and flowering [91]. Their function also consists of regulating the plants' internal clock time in response to changes in light availability throughout the day or season, they ensure that photomorphogenesis and flowering occur during appropriate times in the day or season.

3.3 Interaction of Phytochromes and Cryptochromes with the Circadian Clock and Molecular Mechanisms of Light Signal Transduction

The circadian clock is considered an autonomous system that regulates time-related processes in organisms throughout the day and night. Both phytochromes and cryptochromes are the two photoreceptors that give this clock the required input signals to make plants measure photoperiod accurately (Sanchez et al., 2018). For instance, under long-day conditions, active phytochrome (Pfr) builds up during the later part of the day. CO expression is stabilised by active phytochrome (Pfr) under favourable photoperiods and the CO protein successfully initiates FLOWERING LOCUS T (FT) transcription within the phloem companion cells in the leaves. The mobile signal, often referred to as florigen, is the FT protein, which travels to the shoot apical meristem along the phloem channel through the leaves. In the shoot apical meristem, it encounters other factors (such as FD) to trigger flowering. Pfr → CO expression → FT transcription → FT protein (florigen) transport → flowering induction. Further, cryptochromes regulate other components of the molecular clock to align the expression of genes in the light signalling with the appropriate time. This complex signal-processing system of photoreceptors and clock genes is functionally arranged to effectively analyse light quality, duration and timing into the associated functional programs in plants. The signal transduction of light at the molecular level includes changes in the molecules by phosphorylation and interaction with other proteins. When phytochromes are activated by red light, they move from the cytoplasm to the nucleus, which impacts many transcriptional factors such as Phytochrome Interacting Factors (PIFs). These may either suppress or activate other target genes, deciding the development plan of the plant (Jarad et al., 2020). Like the case with cryptochromes, blue light regulates the stability and activity of several transcription factors to control gene activation and deactivation [45].

3.4 Enhancing Flowering in Rice Through Photoreceptor Manipulation

For a good example of these molecular mechanisms, refer to a recent fascinating study on rice, a short-day plant. CRISPR-Cas9 was used to disrupt the phytochrome and cryptochrome in rice plants to control the CO-FT module of photoperiod sensitivity for desirable flowering time [76]. The results demonstrated that some gain-of-function mutations in phytochromes increase the stability of active Pfr in natural light [29] and hyperactive cryptochrome alleles reinforce the regulation of circadian clocks [19] showed that engineered rice lines with these characteristics flowered later than their wild-type relatives, but without vegetative growth impairment [36]. Experiments with field trial in rice growing areas with seasonal variations in light level [82]) established that through this genetic modification flowering could be induced at an optimum pollination time resulting in an average increase compared to wild-type rice [16]. This study has highlighted the importance of the role of phytochromes and cryptochromes in the process of controlling photoperiodism and demonstrated how we can make use of specific genetic manipulation of these two photoreceptors to make crops grow more efficiently at variable day lengths.

4 Photomorphogenesis: Light-Controlled Developmental Processes

Photomorphogenesis is the process of the perception of light signals and their transformation into stimuli that help regulate the growth and development of plants. In contrast to what is usually known as photosynthesis, photomorphogenesis refers to a package of light-responsive developmental processes during the various phases of plant development. Plants must survive in multiple light conditions, effectively distribute and utilise light, and coordinate developmental stages with external stimuli [51]. Photomorphogenesis is one of the vital processes in plant growth and development, playing a crucial role in converting light signals into architecturally well-controlled plant growth and formation. De-etiolation refers to the morphological change in a dark-grown (etiolated) plant to a light-grown plant type following stimulation by photoreceptors. This is facilitated to allow structural and physiological adaptations like development of chloroplast, leaf expansion and stem thickening adaptions to maximise photosynthesis and the survival of the plants due to light conditions. It is argued that both phytochromes and cryptochromes are significant in this process since they are the photoreceptors that help activate gene expression through complex signalling pathways [60]. Further, the changes in the lateral domain and growth regulation involved in the expansion of the leaf, stem and branching patterns fine-tune the plant architecture. Tomato seeds were grown in controlled conditions (monochromatic light treatments 100% red (R), 100% blue (B), 60% red + 40 per cent blue (RB) and white light (W)) till they germinated in a pot. After they germinated, they were propagated and maintained in sixteen days. The results indicated that red light had a better overall growth than blue light, but a combination of blue and red light (as in RB) prevented excessive elongation and even made it possible to form chlorophyll, encourage the development of photosynthetic apparatus and resulted in a more compact plant structure. Blue light, in particular, induced chlorophyll and carotenoid biosynthesis, stomatal opening and photosynthetic capacity; red + blue light positively affected early developmental parameters and nutrient acquisition. These data indicate that manipulating red green ratios in LED light sources allows growers to manipulate tomato seedling morphology and physiology to fine-tune a balance between vegetative vigour and compactness and shows how photomorphogenic responses to light can be exploited to improve horticultural crop production [31].

4.1 Role of Light in Shaping Early Plant Morphology

Photomorphogenesis is a type of plant development that changes due to light. The seedling de-etiolation is one of the most apparent types of photomorphogenesis. It might best be described in darkness as seedlings with a thin or stretched elongated shape, a well-adapted morphology that glows upwards to its light source. In response to light, the hypocotyl starts to grow downwards, cotyledons pop open and expand and chloroplasts are formed to carry out the process of photosynthesis [78]. These transitions are essential, especially in establishing a good structure that will support the growth of seedlings. Light acts as a signal not only to start the formation of chloroplasts but also to rearrange the cellular organelles and change the metabolic focus towards generating energy and growth.

4.2 Leaf Expansion, Stem Elongation and Branching Patterns

Once the process of de-etiolation starts, light again helps determine the whole plant's architecture. After the seedling stage, photomorphogenic responses are responsible for activities including leaf development, stem growth and production of side branches. It has been revealed that photoperiod and light intensity determine the rate of cell division and the extent of differentiation, from which the size and thickness of the left stem [13]. In one setting, leaves increase in size and thickness to capture more light energy, in high light intensity, the energy capture is higher. In contrast, a plant may grow taller in another setting to overshadow other plants and get to the light. Also, the branching or growth regulation patterns depend on light in a way that the amassing of light at the apex inhibits the growth of lateral buds to focus the plant's energy towards vertical growth. However, in case of low or patchy light intensity, the outgrowth from the lateral bud is favoured, which leads to a bushier structure ideal for maximum light capture.

4.3 Photomorphogenic Modulation in Modern Tomato Cultivars

Tomato leaf and fruit development have demonstrated complex interaction in favour of light quality and plant structure. Tomatoes were produced under two growing conditions using natural light and artificial LED light of specific wavelengths [85]. These changes collectively contributed to the plants grown under LED light, which had increased leaf area size, decreased stem height and increased lateral branching compared to the plants grown under natural light conditions that fluctuated in spectra. These studies showed that the LED treatment modulated the levels of the genes for specific photoreceptors, such as PHYB, which is a large phytochrome, and CRY2 which is critically essential for cryptochrome [2]. This precise regulation led to a strong activation of downstream genes for cell proliferation and, thereby, the generation of plants amenable to high plant population densities and higher yield. This study offers technical advice on controlling photomorphogenesis for crop enhancement and establishes light's significance in plants' shape and physiology (Table 1).

5 Photoreceptors in Photomorphogenesis

Phytochromes and cryptochromes play a critical role in photomorphogenesis by functioning with phototropins and UVR8 to manage various light signals for developmental processes. The plant can perceive a wide range of light due to structural variations and its perception of red, far-red, blue and UV-B light. These entrust the circadian clock using light signals and initiate molecular signalling pathways responsible for seed germination, de-etiolation and flowering. Arabidopsis and rice are recent examples of how genetic engineering of these photoreceptors can enhance plant growth and increase yield; this shows the future of photoreceptor engineering in agriculture [8]. Recent evidence of use of genetic engineering in altering the photoperiod responses to photoreceptors is seen in Arabidopsis and rice. Genome editing in Arabidopsis thaliana with CRISPR-Cas9 and T-DNA insertional mutagenesis has been conducted to generate loss-of-function mutants of alphabetically named phytochromes (e.g. PHYA, PHYB) and cryptochromes (CRY1, CRY2) to gain insight their contribution to the regulation of flowering time. Constitutive promoter-driven overexpression constructs (e.g. CaMV 35S) also have been exploited to boost photoreceptor function and achieve altered transcription of CONSTANS (CO) and hastened or tardy FLOWERING LOCUS T (FT) expression. In Oryza sativa (rice), genetic engineering approaches have been applied to the Hd1-Hd3a pathway by engineering phytochrome (OsPHYB, OsPHYC) genes and cryptochrome (OsCRY1a) genes. With CRISPR-Cas9, gain-of-function alleles have been created stabilising the active Pfr form of phytochrome during normal daylight, hyperactive mutant cryptochromes that modify circadian clock feedback loops, and changes in photoreaction kinetics. Alterations have displayed the capacity to adjust flowering time with

Table 1 Photomorphogenesis and light-controlled developmental processes

Aspect	Description	Significance	Examples	References
Photomorphogenesis	Developmental changes in plants are triggered by light	Essential for adaptation	Optimised growth in varying light conditions	[60]
Phytochromes	Light-sensitive proteins that perceive red and far-red light	Key to regulating growth patterns	Involved in seed germination and flowering	[31]
Light quality	Different wavelengths of light (e.g. red, blue) impact plant behaviour	Influences photosynthetic efficiency, growth and flowering	Wavelengths that match the absorption peaks of key photoreceptors and photosynthetic pigments	[7]
Seedling de-etiolation	The transition from dark to light leads to the development of green leaves	Critical for establishing photosynthesis	Early seedling growth in response to light	[78]
Leaf expansion	Light promotes cell enlargement in leaves	Increases surface area for photosynthesis	Enhanced leaf size under optimal light	[86]
Stem elongation	Stems elongate in low light conditions to reach for light	Improves light capture	Shade avoidance responses in plants	[13]
Branching patterns	Light affects the development of lateral branches	Influences plant architecture	Variations in branching due to light exposure	[85]
Circadian rhythms	Light cues synchronise internal biological clocks	Regulates growth cycles and flowering	Flowering time adjusted by light duration	[2]
Phototropism	Growth towards light sources to maximise absorption	Enhances photosynthesis	Directional growth in response to light	[60]
Shade avoidance	Adaptations in growth when neighbouring plants block light	Ensures survival in dense environments	Increased height in shaded seedlings	[31]
Light-dependent development	Key transitions like germination and flowering are light-regulated	Affects reproductive success	Light sensitivity in flowering plants	[7]
Stress responses	Light interacts with stress signals, modulating plant responses	Enhances survival under stress	Light's role in drought resistance	[78]

(continued)

Table 1 (continued)

Aspect	Description	Significance	Examples	References
Metabolic regulation	Light regulates metabolic pathways affecting growth	Important for energy management	Light-induced changes in chlorophyll synthesis	[86]
Gene expression	Light triggers specific genes that control growth and development	Essential for adaptation	Identification of light-responsive genes	[13]
Environmental interactions	Light responses are influenced by temperature and humidity	Critical for optimising growth	Effects of climate on photomorphogenesis	[85]
Imaging techniques	Advanced methods visualise plant responses to light	Offers insights into cellular behaviour	Use of fluorescence microscopy	[2]
Role of auxins	Hormones modulate growth in response to light	Key for phototropic bending	Auxin distribution changes with light exposure	[37]
Biochemical pathways	Light influences pathways for chlorophyll and other metabolites	Vital for growth and photosynthesis	Regulation of light-harvesting complexes	[2]
Agricultural applications	Knowledge of photomorphogenesis enhances crop management	Improves yield and quality	Controlled light environments in greenhouses	[78]
Future research areas	Identifying gaps in understanding photomorphogenesis	Promotes innovation in plant research	Focus on sustainable agricultural practices	[2]

no detrimental impacts on vegetative growth and it is thus achievable to finely tune crop phenology to variable day length in desired situations. The principles of light perception and signal transduction to support further enhanced crop yield and environmental resilience have risen from various research and technological innovations (Fig. 3).

5.1 Overview of Plant Photoreceptors

5.1.1 Diversity of Photoreceptors

Plants have numerous photoreceptors covering various bands of light wavelengths. They are phytochromes, which are active in red and far-red light; cryptochromes

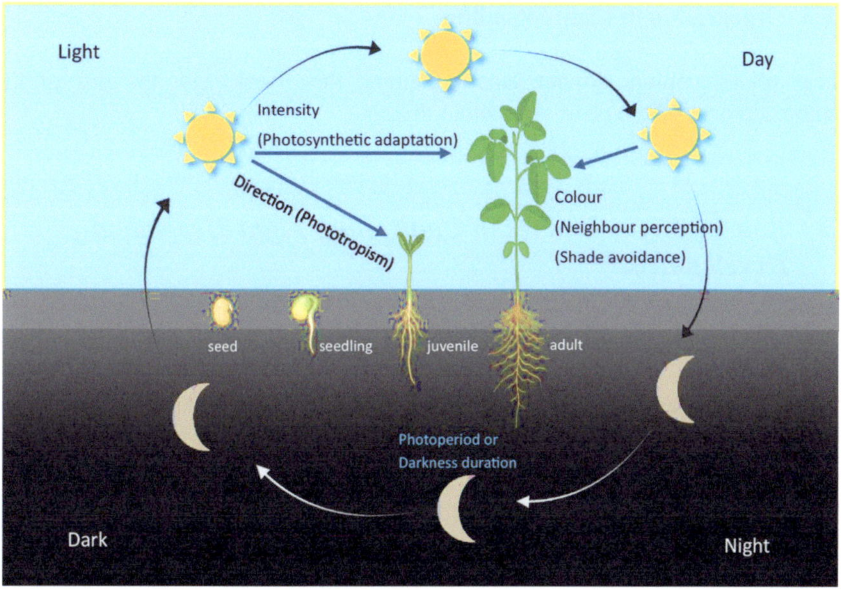

Fig. 3 Dark side of plant growth: a role for phytochrome interacting transcription factors

that are active in blue light; phototropins that move in the direction of the blue light; and lastly, UVR8, active in UV-B radiation [43]. All the photoreceptor classes play a specific role in photomorphogenesis but help the plants develop in response to the light quality and quantity present in the environment.

5.1.2 Functional Specialisation

Phototropins are responsible for directional growth or phototropism, which involves shading plant organs by directing them towards a light source. Phytochromes and cryptochromes regulate developmental processes such as flowering and de-etiolation of seedlings [52]. UVR8, on the other hand, is very important in the plant because it activates stress response mechanisms to shield the plant from possibly destructive UV-B radiation. The exact roles of UV RESISTANCE LOCUS 8 (UVR8) in morphogenesis and growth were mentioned, as well as that UVR8 is a UV-B-specific photoreceptor which triggers photomorphogenic responses through monomerization in response to UV-B and interaction with COP1. This cascade of signalling promotes transcription of genes that participate in flavonoid synthesis, photoprotective pigmentation and prevention of elongation in the hypocotyls, thus affecting leaf expansion and stem structure and general plant shape. UVR8 further mediates stress acclimation which is indirectly favourable to growth under high solar radiation conditions.

5.1.3 Integration in Light Signalling

These photoreceptors do not function alone; they work with the rest of the photoreceptors and the plant's circadian rhythm.

5.2 *Role of Phytochromes in Controlling Light-Dependent Development*

5.2.1 Molecular Architecture and Light Conversion

Phytochromes are photoreceptors composed of biliprotein with a linear tetrapyrrole chromophore. They have two forms that are part of each other: Pr (the red light-absorbing) and Pfr (the far-red light-absorbing). Flavin of red light converts the protein from the Pr form to the Pfr form, which is the biologically active form of the protein that initiates the signalling [10].

5.2.2 Regulation of Developmental Processes

In the active Pfr form, phytochromes move to the nucleus, where they come in touch with transcription factors, especially Phytochrome Interacting Factors (PIFs). This regulation is essential in processes such as seed germination, shade avoidance and flowering time [86]. For instance, in long-day plants, constant activation of phytochrome during the afternoon, constant and active CO, promotes the expression of FT, causing the induction of flowering. Photoperiodic flowering is controlled by an important transcription factor, (CONSTANS (CO)). It is regulated by the circadian clock, but its stability is photoregulated and, under certain conditions of light, is shielded against breakdown by phytochrome and cryptochrome signalling. Stabilised CO levels build up in the phloem companion cells in the afternoon under so-called inductive long-day conditions and directly drives the transcription of the regulator FLOWERING LOCUS T (FT). FT regulates a small and mobile protein, florigen, synthetised in the leaves and then conveyed by the phloem to the shoot apical meristem. There, FT binds with bZIP transcription factor FD to create a complex that induces the expression of floral meristem identity such as APETALA1 (AP1) that initiates the developmental transition between vegetative growth and flowering.

5.2.3 Example in Crop Improvement

A recent study in rice showed that different phytochrome expressions can be controlled using CRISPR-Cas9, and the controlled flowering of rice confirmed this depending on the quality of light. The gene OsPHYC in rice was specifically knock down using CRISPR-Cas9 to investigate the phenotype in OsPHYC and photoperiodic flowering. In PHYC-Loss of-function mutants rice plants in long-day conditions flowered approximately 12 days earlier than in wild-type although vegetative growth remained normal. The modified flowering time was associated with the modified downstream regulators of photoperiod pathway, reduced expression of the floral repressor Ghd7, and stimulation of the CO-FT module stimulating the onset of flowering in response to light quality. The results showed that the transgenic rice plants had better-synchronised flowering, yielding better in a controlled environment agriculture (CEA) [14].

5.3 Role of Cryptochromes in Photoperiodic Responses

5.3.1 Structural Features and Activation

Cryptochromes belong to the flavin adenine dinucleotide (FAD) proteins and, as part of their photolyase/cryptochrome family (PHR) dinucleotide domain, they become responsive to blue and UV-A light [38]. They are evolutionarily related to DNA photolyases-enzymes used to repair UV-induced DNA damage, and have since acquired a photoreceptor role and regulate photomorphogenesis, circadian rhythms and flowering. Conversely, phytochromes have no evolutionary connection with DNA repair, and only perform the role of red/far-red light photoreceptors cryptochromes act as signal receptors that control blue light perception.

5.3.2 Influence on Circadian Regulation

Cryptochromes regulate circadian rhythm by synthesising clock genes due to their influence on other proteins and molecules in living organisms [74]. They play a role in regulating gene oscillations, thereby providing the required synchronisation to physiological activities such as the opening of stomata, expansion of leaves and light-controlled morphogenesis, amongst others, during light and darkness.

5.3.3 Photomorphogenic Impacts

Besides regulating circadian rhythm, cryptochromes have a direct impact on plant morphology, especially growth [75]. Their activity suppresses hypocotyl elongation and stimulates cotyledon enlargement during de-etiolation, which helps the transition from dark to light-grown developmental processes in tomato [90].

5.4 Phototropins and UVR8: Mediators of Phototropism and UV Response

5.4.1 Phototropins—Directional Growth

Phototropins are essentially blue light receptor proteins that play a significant role in determining phototropism, which is the bending of a plant towards the light source. They possess LOV, an acronym for light oxygen voltage, and these promote the sensation of blue light, leading to the reorganisation of the cytoskeleton that directs the elongation of cells in a given direction [79]. This response is essential since it helps increase the chances of capturing light by bending its parts towards the light source, the leaves and stems.

5.4.2 UVR8—UV-B Perception and Stress Responses

UVR8 is a photoreceptor specialising in detecting UV-B light radiation. Upon exposure to UV-B, UVR8 changes from its dimeric form to the monomeric form, which triggers several signalling pathways that lead to the expression of protecting genes, including those that produce enzymes involved in the synthesis of flavonoids [23]. These compounds also protect plant tissues from UV injuries and induce stress signalling, improving plant stress tolerance.

5.4.3 Integration with the Circadian Clock and Molecular Signal Transduction

Phytochromes and cryptochromes are related to the circadian clock in many aspects. These interactions ensure that the light signals are synchronised with other internal timing systems in the plant to match developmental processes. For instance, the principal CO-FT module involved in photoperiodic flowering is controlled by light signals that are perceived through these receptors, which do so by entraining the circadian clocks [25]. The reasonably complex signalling includes protein phosphorylation and destruction of negative regulators such as PIFs and some of the proteins that lead to the expression of light-responsive genes.

5.4.4 Enhancing Crop Yield Through Photoreceptor Engineering in Arabidopsis and Rice

Genetic engineering of target modulation of phytochromes (red/far-red light), cryptochromes (blue/UV-A light) and phototropins (blue light in phototropism and stomata opening) coupled with precision light regimes can tune photomorphogenesis, flowering time and photosynthetic capacity/efficiency in Arabidopsis or rice. This concerted manipulation maximises light interception, biomass production and reproduction timing and enables the grain yield to be maximised in variable photoperiod and light-quality conditions. In Arabidopsis, a hyper-stable CRY2 and a stabilised PHYB were generated to enhance de-etiolation and biomass production by entraining short-day signals under fluctuating light. At the same time, changes in rice gene phototropin resulted in improved phototropism, allowing the rice plant to adjust the angle of the leaves for maximum light intensity [63]. Those on the field trials showed that apart from flowering at the right time, grain yield increased by 15%. This fact testifies to the high potential of photoreceptor engineering to increase the efficiency of plant cultivation when the plants use objective surplus illuminations with increased brightness, simultaneously improving their circadian rhythm, which, in turn, suggests the possibility of further developing light stimuli regulation in plants.

6 Hormonal Control of Photomorphogenesis

From basic photomorphogenesis, many aspects are regulated by plant hormones, which are generally instants in plant development. They cooperate with other signalling systems, such as light signal transductions, to regulate growth and developmental processes. These include gibberellin, auxin and brass-steroids, which are responsible for different functions in the aspect of light sensitivity. This paper further discusses how these hormones interact with light and also explains in detail how hormones control growth in response to these lights, and then presents a case on how gibberellins regulate hypocotyl elongation under different lights with new scientific and versatile language to explain this somewhat complex processes. Thus, the interrelationship between light and hormonal signals constitutes one of the realms of photomorphogenesis. Principal photoreceptors, including phytochromes and cryptochromes, regulate hormonal pathways as in the most important hormones, gibberellins, auxins and brassinosteroids to control developmental switching from elongation in the darkness to vigorous growth in conditions of light [37]. The Arabidopsis example sheds light on how model organism is integrated by light to control hypocotyl elongation, as the new example of rice reveals how hormonal alteration may be combined with light signalling to improve crops [26]. Altogether, these findings provide bases for new approaches in plant growth and production to environmental changes, thus establishing the standard form of plant output and efficiency.

6.1 Interaction Between Light Signalling and Plant Hormones

6.1.1 Gibberellins (GAs), Auxins and Brassinosteroids as Modulators of Light Responses

Light and hormones are not simple messengers; they are tightly integrated to convey information necessary for the correct development of plants. Gibberellins (GAs) play the chief role in the elongation of cells and cell division. At such high levels of GA, the hypocotyl grows longer, helping a seedling to come out of the soil in the dark [3]. However, when light is perceived, photoreceptors such as phytochromes negatively affect GA biosynthesis or up-regulate the GA catabolism, inhibiting shoot growth and the onset of the de-etiolation process. Auxins derived from Bryant's model which hypothesises that asymmetric distribution of auxin between light and dark sides of plant tissue contributes to the directionally non-uniform expansion of plant tissues–well-known to mediate cell elongation in plants. This is regulated by light via their polar transport domains, such that plants can align their growth to environmental responses. The growth could be directed up or away again as on different sides of the stem it causes redistribution and mobilisation of the auxins on a light signal and causes the seedling to curve towards or away the light source. For instance, in phototropic bending, less auxin accumulates on the side of the stem facing the light than on the shaded side, thus resulting in differential rates of cell expansion [84]. BRs interact with both GAs and auxins to influence growth. BRs are required for well-being and have been associated with regulating light-mediated response to adjust cell growth and differentiation. BR signalling can be fine-tuned under light conditions, allowing plants to coordinate cell expansion, photomorphogenesis and stress adaptation according to the prevailing light environment. Other hormones' growth inhibitory effects can be negated so that the cells can expand optimally in response to the available light-scarce resources.

6.1.2 Integration of Hormonal and Light Pathways

Light and hormone-activated receptor molecules are points of convergence that are implemented through transcription factors at the molecular level. Some of these proteins include DELLA repressors that regulate GA signalling in response to other components of phytochromes. In this case, in the dark, DELLA proteins are degraded, thereby allowing elongation in the presence of GAs to take place, however, in the light, DELLA proteins are stabilised from degradation and restrain GA-promoted elongation [70]. Similarly, some proteins that regulate genes involving auxin include the auxin response factors (ARFs) regulated by light-activated transcription factors and, therefore, cause temporal and spatial changes that direct developmental processes [34]. Thus, in addition to regulating gene expression on its own, components of the brassinosteroids signalling also interact with elements that involve light [89]. Therefore, coupled with the control of gene expression itself, the brassinosteroid signalling system interacts with pathways of light in several ways through the common regulators that include BIN2 kinase and PIF transcription factors. Phytochromes and cryptochromes, which are photoactivated, regulate the synthesis and responsiveness of BRS and negative regulators of photomorphogenesis such as BZR1 antagonise the light repressive effect of HY5. Such crosstalk is elaborate to match the plant's erect pattern as per current light conditions surrounding it.

6.2 Hormonal Regulation of Light-Dependent Growth and Development

6.2.1 Dynamic Hormonal Balances and Growth Adaptation

The hormonal and the light signals are two orchestras that control the switch between elongation and differentiation. The process of de-etiolation in seedlings involves some drastic changes to their hormonal profiles. During germination, high levels of gibberellin (GA) provide prompt embryo growth to allow seedlings to make a fast transition into light-dependent development. Along with the process of photoreception, significant changes happen when seeds are exposed to light, namely, the induction of photoreceptors, DELLA proteins degradation, stimulating cell elongation, the production of chlorophyll, activation of photomorphogenesis. GA levels are reduced; DELLA proteins stabilise; cell elongation is the eventual inhibition [71]. At the same time, auxin distribution assists in the enlargement of the cotyledons and the formation of chloroplasts so that photosynthesis can occur. BRs regulate growth via cell wall loosening and a constant moderate level of cell expansion but at a breakneck pace to meet all the requirements [62].

6.2.2 Feedback Mechanisms and Environmental Sensing

These hormonal adjustments are not fixed; they are part of a loop of endocrine communication that receives an evaluation of the light environment in quality and quantity. For example, photoreceptor activation changes the expression of genes that encode regulatory proteins in the biosynthesis and degradation of hormones [12]. Such feedback loops help create the best timing for the plant to develop about its surroundings. This cyclic regulation helps the plants integrate with day and night environments and changes in light periods in a year, defining their flowering, stress-managing ability and other factors contributing to plasticity.

6.2.3 The Role of Gibberellins in Hypocotyl Elongation

In a study in Arabidopsis, researchers discovered how gibberellins control hypocotyl growth under different light conditions [9]. In the seedlings growing in the dark, high levels of GA facilitate the elongation of the seedling's hypocotyl, which is essential when it emerges from the ground. However, when seedlings are exposed to red light, phytochromes to their active Pfr form and a signalling pathway is generated, leading to the deactivation of the GA biosynthesis enzyme and activation of DELLA. This hormonal change results in a drastic reduction in the hypocotyl growth and the beginning of the de-etiolation process. In an innovative attempt, they altered levels of GA by applying exogenous gibberellin and using plants with genetically modified sensitivity to GA [35]. What was found remarkably is that the combination of moderate red light and small changes of GA can somewhat rescue the elongation. In contrast, high-intensity light almost counteracts GA effects due to extreme DELLA accumulation. This paper underlines the importance of a delicate equilibrium between GA signalling and light perception and how the latter is essential in regulating the change from dark to light growth. The information gathered from this study is very relevant to crop enhancement since hypocotyl elongation directly impacts seedling growth and early survival, especially in fluctuating light conditions.

6.3 Integrative Hormonal Manipulation in Crop Improvement

6.3.1 Advancing Agricultural Productivity Through Hormonal Modulation

An example of rice, considered below, indicates that the integrated approach in hormonal control and regulating light signals effectively enhances plant growth results. Scientists have proven that genetic engineering can be used to control the biosynthesis of GA in rice plants for specific light conditions using CRISPR-Cas9 [80]. Therefore, after manipulating the regulatory genes in the GA pathway, the function of the modified rice plants was to have appropriate hypocotyl length depending on whether the environment was shaded or full sun to enhance seedling vigour and subsequent biomass formation. Observations from the field tests deduced that these plants had better early vigour, higher tolerance to light fluctuations and higher yield rates. Thus, the given example describes the use of hormonal control in photomorphogenesis and further shows how the principles could be implemented to solve the existing problems of agriculture (Table 2).

Table 2 Hormones interact with light signalling to regulate photomorphogenesis and plant development

Hormone	Role in photomorphogenesis	Interaction with light signalling	Effect on plant growth and development	Example of response	References
Gibberellins (GA)	Promotes stem elongation	Regulated by light intensity quality	Stimulates germination hypocotyl growth	Seed germination in darkness	[37]
	Inhibits photomorphogenesis	Degraded in light	Shorter hypocotyl in light-grown plants	Shade-avoidance response	
Auxins (IAA)	Regulates cell elongation	Light inhibits auxin transport	Differential growth in seedlings	Phototropic bending towards the light	[3]
	Controls apical dominance	Interacts with phytochrome signalling	Root and shoot development	Enhanced shoot elongation in low light	
Brassinosteroids	Promotes cell expansion division	Downregulated by light exposure	Stimulates elongation in darkness	Skotomorphogenesis (dark growth)	[84]
	Enhances photosynthetic efficiency	Works with auxins in shade responses	Increases leaf expansion in low light	Shade-induced stem elongation	
Abscisic acid (ABA)	Inhibits seed germination	Modulated by light through phytochromes	Controls stress responses	Delayed germination under light exposure	[70]
	Regulates stomatal closure	Reduced in high-light conditions	Protects against excessive transpiration	Stomatal closure under drought stress	
Cytokinins (CK)	Promotes cell division	Interacts with light for leaf expansion	Delays leaf senescence	Increased chlorophyll retention in light	[71]
	Stimulates chloroplast development	Upregulated by light	Enhances photosynthesis	Greening of etiolated seedlings	
Ethylene	Inhibits elongation in light	Antagonistic to GA and auxins	Controls seedling growth	Triple response in dark-grown seedlings	[12]

(continued)

Table 2 (continued)

Hormone	Role in photomorphogenesis	Interaction with light signalling	Effect on plant growth and development	Example of response	References
	Enhances stress responses	Modulated by light quality	Promotes leaf senescence	Leaf abscission under stress	
Jasmonic acid (JA)	Mediates stress responses	Induced by UV-B light	Regulates defence mechanisms	Increased secondary metabolite production	[15]
	Suppresses hypocotyl elongation	Works with phytochrome pathways	Inhibits excessive stem growth	Reduced elongation under high-light	
Salicylic acid (SA)	Activates plant defence genes	Modulated by blue light	Enhances pathogen resistance	Increased resistance under UV exposure	[80]
	Promotes chlorophyll accumulation	Works with cytokinin pathways	Enhances photosynthetic efficiency	Delay of leaf senescence	
Strigolactone	Regulate shoot branching	Inhibited by high-light intensity	Controls resource allocation	Reduced shoot branching in full sunlight	[3]
	Mediates hormonal crosstalk	Works with auxins in shade avoidance	Suppresses lateral bud growth	Enhanced shoot elongation under shading	
Melatonin	Protects against oxidative stress	Light regulates its biosynthesis	Enhances stress tolerance	Antioxidant defence in high-light stress	[71]
	Delays leaf senescence	Works with cytokinins ABA	Enhances longevity of plant tissues	Delayed yellowing of leaves in intense light	
Polyamines	Promote cell division growth	Regulated by light conditions	Protects against oxidative damage	Improved seedling vigor under high light	[37]
	Modulate hormone interactions	Interact with auxins cytokinins	Enhances chlorophyll synthesis	Increased photosynthetic capacity	
Karrikins	Stimulate seed germination	Perceived through light signalling	Promote early seedling growth	Enhanced germination after fire exposure	[80]

7 Environmental Factors Affecting Light Perception

This section demonstrates in light perception among various plants and utilisation of light are influenced by various developmental factors, including light intensity or quantum, spectral quality, photoperiodism and other environmental factors such as temperature and water availability [12]. All these factors are important in regulating plant growth and development by affecting the activity of photoreceptors, hormones and genes. The new example in Mediterranean tomato breeding shows that achieving light integration with other measures means producers of environmental protection can be reached, and thus is signifying a positive way through which crop improvement can be done sustainably amidst climate change.

7.1 Impact of Light Intensity, Quality and Photoperiod on Plant Growth

7.1.1 Light Intensity and Its Effects

Light intensity is one of the most significant parameters influencing the rate of photosynthesis and plant growth. Stimulation of carbohydrate production occurs when light intensity is moderate since the chlorophyll molecules are often excited by light absorption. However, the high light intensity can cause photoinhibition, where damage to the photosynthetic machinery occurs due to the generation of high reactive oxygen species (ROS). On the other hand, low light conditions slow the flow of energy and cause conditions such as etiolation, which features elongated stems, small leaves and low chlorophyll content. Consequently, plants can change and regulate the size, shape and area of the leaves, the frequency of opening stomata and the photosynthetic pigments used to capture light to minimise the impact of light intensity [44].

7.1.2 Light Quality and Spectrum Composition

The relative amount of red, blue, far-red and ultraviolet light is the most important feature of the condition in which plants grow. For instance, phytochromes sensitive to red light are needed for the appropriate developmental processes such as germination and flowering. Blue light perceived through cryptochromes and phototropins is essential in photomorphogenesis, including stomatal opening, phototropism and inhibition of hypocotyl elongation [66]. It helps regulate shade avoidance, the presence of the neighbours and ultraviolet or UV-B light triggers the synthesis of protectant compounds, especially sunscreens, especially flavonoids.

7.1.3 Photoperiod and Seasonal Adaptation

Photoperiod is the number of hours of light within a day, which affects flowering time, dormancy and other plant developmental stages. Regarding photoperiod, the plants are classified into short-day, long-day and day-neutral [4]. These classifications clearly illustrate that light duration and hormonal regulation are intertwined to decide the time developmental processes occur in plants. Cycling their processes in phase with the light conditions, plants regulate the expenditure of energies expended towards causes such as reproduction to preferred times.

7.2 *Effects of Different Wavelengths (Red, Blue, Far-Red, UV) on Plant Development*

7.2.1 Red and Far-Red Light

Red light, particularly with a wavelength of approximately 660 nm, effectively excites the phytochrome receptors in two high- and low-energy states. The effect of light on germination, flowering and other such photomorphogenic responses results from the photoconversion of the inactive Pr form into the active form known as the Pfr form after absorption of red light [81]. The phytochrome system is controlled by this far-red light (about 730 nm), which realigns it to the inactive form. Therefore, the ratio of red to far-red light also indicates the light conditions in which higher values promote photomorphogenesis while low, indicated by shadowing, bring about responses such as elongation of stems.

7.2.2 Blue Light

Blue light with a wavelength near 450 nm is sensed by the cryptochromes and phototropins. Cryptochromes control the growth inhibitory effect observed in hypocotyls and play a role in the entrainment of the circadian rhythms to the light–dark cycle so that the growth and development of plants coincide with the periods of day and night, respectively, [75]. Phototropin is the protein that plays a role in phototropism, in which the plant bends towards the light due to differential expansion of the cells on one side. This reduces the amount of wasted light while enhancing how the plant uses the available light to perform photosynthesis.

7.2.3 Ultraviolet (UV) Light

Despite being a small portion of the sunlight, UV-B radiation has many effects on living organisms. Through activating its downstream signalling, the UVR8 photoreceptor involved in UV-B perception is associated with the biosynthesis of protective compounds such as flavonoids and phenolic compounds [69]. These compounds interfere with UV-related destruction by either intermingling or chipping the hazardous UV radiation hence hindering its infiltration into the tissues and consequently damaging the molecules and deal with ROS. Aside from the protective systems, UV-B radiation can also adjust developmental events like regulation of stomata development for transpiration processes, stress response and acclimation for changes in photic levels within the environment.

7.2.4 Optimising Growth in Controlled Environment Agriculture

Field research conducted on a modern vertical farm in the Netherlands reveals the variation in the impact of the light's quality on plant growth [35]. The article's authors experimentally controlled LED lighting systems to produce a spectrum specific for lettuce production. With precise control of the intensities of red, blue and far-red light, the study thus provides excellent manipulation of plant morphology. Generally, high blue light ensures compact growth of plants due to inhibition of stem extension and causes bright leaf pigments to be produced, including anthocyanins, which leads to healthier and more colourful leaves while high red light promotes biomasses [67]. Furthermore, exposure to a small amount of UV- B light increases the production of antioxidants, hence improving the nutritional value of the crops. This controlled environment also enabled an understanding of the manipulation of various light wavelengths to get better yield and quality in plants in agricultural settings.

7.3 Interaction of Light with Other Environmental Cues

7.3.1 Synergistic Effects of Light and Temperature

Temperature is an important factor in the external environment that has a direct correlation between its effect and light to control plant growth. So, it can be accepted that an optimal temperature level improves the effectiveness of light-dependent phenomena, like energy production in photosynthesis and enzymes for metabolic pathways. However, low or high temperatures can affect these processes to a great extent. For instance, higher temperatures may worsen some effects of bright light by speeding up the photoinhibition rate; on the other hand, low temperatures can slow down some effects of light linked to the developmental process [47]. Light and temperature signalling pathways are interconnected, and both signals activate necessary genes through specific transcription factors and hormonal controls to guarantee proper plant growth according to the weather conditions.

7.3.2 Influence of Water Availability

Water availability is another factor affecting the plant interaction with light. Water in adequate quantity is crucial because it helps in cells' turgidity while playing a vital role in photosynthesis [5]. In response to water deficit, other activities like cell division of leaves and stomatal movements are downregulated, and even photosynthesis is reduced. Also, water deficiency affects nutrient and signal molecule movements, leading to some photomorphogenic responses. Other well-functioning mechanisms are, for instance the maximising of a surface-to-volume ratio through changing leaf size and shape so that light can be caught to the maximum with water being lost in the minimal measure. Also, modifications in leaf angles (e.g. less horizontal) eliminate the problem of excessive light being absorbed in excess under high irradiance, thus avoiding photo-damage. The thickening of the leaves, especially palisade mesophyll, increases the density of chloroplast and the rate of photosynthesis thus enabling the leaves to absorb light and modify the energy more adequately. In combination, these morphological adaptations facilitate light harvesting to compensate protection, resulting in better growth and stress tolerance of plants in general.

7.3.3 Integrated Environmental Sensing

Light, temperature and water are the aspects, that a plant must sense for its existence, success and reproduction. These signalling carriers include phytochromes, cryptochromes and hormones, through which plants can capture multiple environmental signals and change their growth and development patterns in response [65].

7.3.4 Climate-Resilient Crops in Mediterranean Regions

In a recent effort to improve the performance of its crops during one of the harshest climate disasters, researchers in Spain genetically modified tomato varieties to perform optimally under Mediterranean growing conditions where high light intensity, high temperatures and water stress are normalcy. By improving the signals from the light receptors and the drought-responsive hormones, the new tomatoes sustained a high yield during the typically unfavourable microclimate for growth [1]. In field trials, it was established that the photosynthetic and water-use efficiency in these tomatoes was higher, further proving that light signalling must be interconnected with other environmental signals for the crop to perform better [92]. Some genes were catalysed to be expressed and in further effect, downstream stress response genes were activated and the evolution of growth-regulating hormones like abscisic acid (ABA) and gibberellins ensured that they could constantly change according to the conditions of the environment.

8 Applications in Agriculture: Manipulating Light for Optimal Growth

The regulation of light as a factor influencing and promoting plant growth has become one of the fundamental preconditions of today's agriculture. Thus, not only is significant in regulating photosynthesis and photomorphogenesis processes defined by light quality, intensity and duration but all the practical aspects of crop scheduling and yield improvement [53]. Recent advances in controlled environment agriculture and genetic breeding have unlocked new opportunities to manipulate photoperiodism, refine indoor lighting strategies and enhance seedling quality—thereby paving the way for improved crop performance in both open-field and indoor farming systems. Manipulating light for optimal growth in agriculture offers multifaceted benefits that range from precise control of photoperiodism and flowering time to enhanced plant architecture through controlled environment agriculture. With intelligent darkness control, dynamic lighting control and genetic modification of plants for changed photoreceptor sensitivity, LED lighting technology is changing cropping frequency and quantity improvement. The introduction of new applications in major light and genetic technologies has led to better germination quality of lettuce seedlings under the vertical farming system, better uniformity and growth rates. In rice, CRISPR-Cas9 has been used to manipulate photoreceptor gene-such as OsPHYC in *Oryza sativa* indica varieties to generate lines with an optimised floral induction time to different photoperiodic environments to enhance yield stability [94]. These instances show the benefits of combining genetic instruments and precision lighting to enhance crop growth productivity in the controlled setting and in the field. When it comes to the perception of light and its regulation in photomorphogenesis, the combination of environmental and genetic practices will be fundamental when approaching the problems in global agriculture and food production in the future.

8.1 Practical Implications of Photoperiodism in Crop Scheduling

Photoperiodism, the effect of the length of the day on the plant's activity, is a decisive factor that sets the timetable for flowering, fruiting and harvest. Knowing the photoperiod requirement is fundamental in determining crops' planting and harvesting periods in a region. For instance, some plants, like short-day plants, need long periods of darkness to induce flowering, whereas others, like long-day plants, need short periods of darkness. For instance, growers may use supplementary light or blackouts to control the photoperiod.

8.1.1 Control of Flowering and Harvest Time

In a controlled environment, manipulation of light duration allows growers to bring the flowering time forward or set back to meet the demand in the market. For instance, in a greenhouse setting, the day length can be elongated through supplemental LED lighting. This would enable the operators to continue harvesting flowers throughout winter when the flowers would generally take time to bloom [64]. It also ensures that yield is favoured by controlling the seasonal changes to suit production requirements and not vice versa.

8.2 Controlled Environment Agriculture: Using Artificial Lighting to Regulate Growth

More progress in using LEDs in this area has made indoor farming possible due to its efficiency and controlled light spectrums that can be fashioned to emulate those of natural sunlight. LEDs allow growers to regulate the quality (wavelength and intensity), and duration of light, which in turn impacts growth at a cellular level. This technology is important in controlled environment agriculture (CEA), where environmental elements such as temperature, moisture and nutrient solution are controlled, in addition to light.

8.2.1 Role of LED Lighting in Indoor Farming

LEDs offer precise wavelengths of light that can be used to specifically promote photosynthesis and photomorphogenic regulatory mechanisms, and in that way maximise plant growth and development. Blue light helps the plant's elongation and enhances the leaf's thinness, whereas red light helps in the flowering and forming biomass [86]. Variation of the ratio of these blue and red wavelengths means that growers can control plant architecture to improve the quality and homogeneity of the products. Moreover, given that LEDs are long-lived devices that emit little heat and have low power consumption, the operational expenses and energy losses remain low when used for indoor farming systems [61].

8.3 Strategies for Enhancing Photomorphogenesis in Crops and Breeding for Altered Light Sensitivity

Enhancing photomorphogenesis can lead to the development of more potent and healthier seedlings with higher photosynthetic rates, which may further result into better and improved yield [49]. The following approaches can be used to improve plant response to photomorphogenesis a spectrum of light exposure during seedling growth and gene manipulation to alter a plant's inherent photosensitivity [59].

8.3.1 Improving Seedling Quality Through Light Manipulation

Scientists cultivated lettuce sprouts in the light emitted by LED lamps tuned to wavelengths of blue and red colours. The blue light was initially enhanced during the beginning stages to develop erect, massive, dark green leaves and a strong root system. Once the seedling morphology was established, the light spectrum was changed to red light to set other conditions, such as biomass production and flowering. Compared with the natural sunlight or general white light-grown seedlings, the plants grown under the designed LED spectrum light showed enhanced homogeneity and higher chlorophyll levels [59]. This skewed photomorphogenesis study underlines the role of dynamic light in increasing the yield and growth of plants in commercial fields.

8.3.2 Breeding Crops with Altered Light Sensitivity for Improved Yield

In addition to changes in light exposure to the external environment, selective plant breeding and genetic modification of organisms also enable internal modifications of a plant's light responses. Thus, by selecting for or engineering variants that altered the expression of genes for photoreceptors, such as phytochromes and cryptochromes, the breeders developed cultivars that were more amenable to different light environments. As an example, some important allelic variations involved in influencing flowering time under short-day conditions have been discerned in rice phyto-genetics. Remarkably, mutations in the Hd1 gene, a homolog of Arabidopsis CONSTANS (CO), change the photoperiod sensitivity, regulates the expression of the rice analogue of FLOWERING LOCUS T (FT), the Hd3a. Other floral repressors of great importance are variants of Ghd7, which delay flowering when plants grow in the long-day conditions but promote early flowering under short-day conditions. As well, variant of Ehd1 and RFT1 genes contribute to the time of tune the time of flowering. Such genetic adaptation allows rice cultivars to accommodate flowering to fit various settings, at the maximum profitable yield and completion of lifecycle [54]. With the help of marker assisted selection (MAS), some of the most significant molecular markers associated with flowering time genes have been employed to introgress these alleles in high-yielding rice varieties including simple sequence repeats (SSRs), single nucleotide polymorphism (SNPs) which are related to the Hd1, Ghd7, Ehd1 and RFT1 genes. This has made controlled flowering to ensure synchronised flowering is possible and hence crops are able to make optimal use of the available long growing season hence increasing yield stability and flexibility of crops to adapt to various agro-climatic zones. This genetic approach works hand in hand with environmental changes and is a good way to enhance the growth of crops in the climate change era.

9 Climate Change and Its Impact on Light Perception

The natural environments of light are being changed by climate change due to the change of seasonal patterns, cloudiness and weather patterns that adjust the interpretation of day length and light quality. These alterations interfere with plant photoperiodism-the process through which plants quantify the length of day to time sensitive growth and development factors such as flowering and dormancy. Disrupted photoperiod patterns bring about predictive flowering, a smaller reproductive performance and inability to synchronise with pollinators or optimal growth conditions. This reduces plant growth, productivity and ecology. It has become even more important to understand these effects in order to come up with more climate tolerant crops and climate flexible farming strategies that will ensure sustainability of agricultural production in response to annual changes in photoperiods due to climate change. Varying day lengths, mainly due to shifts in seasons, coupled with the increasing effects of artificial light pollution on the world, are some of the areas that affect the adaptability of plants [50]. Consequently, there are significant challenges in agriculture and the sustainability of ecosystems, and it is necessary to develop new approaches to managing the photoperiodism of plants under altering light conditions [59]. Climate change and the global growth of artificial light interfere with plant photoperiodism in multiple ways, ranging from flowering time to overall fitness. Variations in day-night duration affect the coordination of developmental activities, altered light conditions affect the circadian cycles and factors that negatively impact yields in agriculture and the stability of the ecosystem [22]. In anticipation of these challenges, it will be important to embark on an integrative approach that relies on genetic engineering to adapt to the cultivation of crops that would be responsive to a rapidly changing environment as far as the light factor is concerned. The case in the European Alps reveals the possibility of such strategies, reflecting that much more drastic changes are needed to pursue sustainable agriculture that would sustain the changing climate of the global environment.

9.1 Effects of Shifting Photoperiods Due to Climate Change

Seasonal variations are shifting in a manner that changes the onset, duration and extent of daylight exposure in different latitudes. Historically, plants depend on specific day lengths to control their photoperiodic developmental stages, including flowering and dormancy. However, the earth's global temperature increases and changes in weather conditions affect these photoperiodic signals. For instance, increased occurrences in springs and longer growing seasons may alter the onset of flowering, disrupting the relationship of plants with their pollinators and exposing them to frost. This may lower reproductive success and decrease yield, thus suggesting that there is a direction of geometric algebraic misalignment for plants [20]. Besides, photoperiod affects the degree of proportionality between the growth of the vegetative and generative organs of the plant. Disruptions in the timing of light exposure mean that the intricate regulation systems of phytochromes and cryptochromes could be disrupted, thus, there will be poor architecture and allocation of resources. Studies in the temperate climate zone have postulated that changes brought by variation in photoperiods may bring about early flowering in some plants and delay maturation in others in case of extended daylight during critical development periods [58]. These disruptions demonstrate how plants respond to environmental alterations that occur in a very short time.

9.2 Impact of Global Light Pollution on Plant Photoperiodism

Similarly, and in consonance with climatic change, escalating global light pollution also hinders the basic natural photoperiodic cues relied on by plants. That streetlights and other artificial lighting from urban areas bring forward the artificial dusk to alter the phenomenon called 'skyglow'. For instance, artificial moonlight in the sky is constantly present, which can disrupt correct night sensing that is important in triggering processes such as flowering, seed germination and synchronising the body's internal clock [50]. In many ecosystems, the phenology of plants in or adjacent to the urban environments is different, for instance, short-day plants may take an unusually long time to flower due to artificially prolonged day length. These interferences can cause chain reactions in ecological processes—changing the competition between species and synchronising food waves. For instance, light pollution can worsen the adverse impacts of climate change since it interferes with circadian rhythms and stress, adversely affecting plants' fitness. Suburbanisation of urban habitats threatens to worsen due to progressive urban expansion. Hence, there is a high chance that natural and artificial light conditions will probably merge, thus harmfully influencing the lives of species living in harsh environments in the production of crops.

9.3 Future Challenges: Adapting Crops to Changing Light Conditions

One of the most striking yet probably one of the biggest problems is creating plants capable of withstanding two factors—photoperiod and lighting conditions. Opportunities to develop cultivars belonging to low-light liberal species by applying plant breeding and genetic engineering are now numerous. For example, selectively altering genes of photoreceptors like phytochromes, cryptochromes and phototropins can help adapt the developmental processes of plants to specific changes in brightness [39]. Also, in controlled environment agriculture (CEA), when using LED lighting systems, it is possible to replicate those desirable photoperiods and prevent the negative impacts of environmental fluctuations. Thus, by controlling the quality of light, its brightness and time exposure, CEA can ensure stable crop productivity despite climatic fluctuations outside.

9.3.1 Adapting Alpine Crops in the Face of Climate Change

One relevant example of such challenges and solutions was evidenced in a study done in the European Alps [18]. These alpine areas, previously having short season lengths and different photoperiods, now have longer days due to earlier snow melting and increased temperature. [6], noted that local farmers have observed early flowering and maturation of traditional crops such as barley and potatoes that are being grown, resulting in the wrong timing of harvesting periods and high susceptibility to late frosts. As a result, several workers conducted studies with agronomists to work on new crop varieties with different light sensitivity. By down or upregulating the genes involved in the photoperiod pathway using CRISPR-Cas9, the team successfully developed the barley lines that continued growing and flowering on time even when grown under long-day conditions [41]. Actual tests verified whether such crops deliver sustained yields and improved outsized vulnerability to climate fluctuations, these rendered the whole concept a relatively viable approach for the affected areas [11].

10 Conclusion

Photoperiodism and photomorphogenesis are fundamental processes in how plants manage and are sensitive to light to put various processes in motion. In this chapter, several findings pointed out the various pathways involving phytochromes, cryptochromes, phototropins and UVR8 in perceiving the light signals from the environment and the response the cell would give to such light stimuli. Phytochromes response to red and far-red light are involved in critical processes, including seed germination, shade perception and flowering. These are mainly sensitive to blue light, and apart from regulating circadian rhythms, they play a role in phosphorylation, regulating growth by promoting stem and leaf expansionshrinkage. Phototropins are involved in phototropism since they regulate the growth orientation towards the light source to enhance photosynthesis. At the same time, UVR8 recognises UV-B radiation to initiate protective mechanisms that protect the cell from damage from deleterious light. Study of these light-mediated factors is highly important to elucidate the working and regulation of plant development, particularly in the natural environment and under agricultural conditions. According to photoperiodism and photomorphogenesis, plants can finely regulate their growth, reproduction and stress response according to their natural environment's daily and annual light patterns. This is important in species perpetuation, community functioning and productivity for species' existence and management. Such knowledge enables farmers to have the best crop output when grown in different and sometimes unfavourable light conditions. Thereby, the benefit of crops can be tweaked to exhibit vigorous growth, Oxidative Stress Index (OSI), stress resilience and production reliability. For instance, modifying phytochrome adaptation may resolve issues with flowering and fruiting to coincide with adequate lighting while optimising cryptochrome responsiveness, which may see the plant growing in compact structures to accommodate the light system. The following research is expected to investigate the prospects of genetic engineering of light signalling as a yardstick for increasing the yield of crops. New genome editing techniques, most notably CRISPR/Cas9, have created new possibilities to work with genes related to light perception and transmission mechanisms. Thus, when the defined photoreceptor genes or their downstream effectors are manipulated, creating new crop varieties that respond to a specific environmental light signal becomes easy. Some of these variations could increase ability to effect change at specific favourable wavelengths, hence increased growth rate or yield and stress resilience under light intensity. In addition, the application of omics approaches together with computational modelling will provide a better understanding of the specific nodes in the light signalling and intricate interactions between different pathways, which will be helpful for the engineering of the synthetic systems to optimise preferable traits. This is supported by collaborations of scientists/departments of molecular biology, agronomy and data science/analytics to translate the research into practical tools for agriculture. The need of having such efforts in ensuring the development of new varieties of crops that can perform equally well in the controlled environment of tests and the field considering the onslaught of climate change. In conclusion, knowledge of photoperiodism and photomorphogenesis will help develop sustainable ways of practising agriculture that will help feed the population and optimise the use of the environment. Further studies into this would directly increase crop output, which would eventually promote improved precision agriculture and improved and strong plant types to support the growing population.

References

1. Abbas K, Li J, Gong B, Lu Y, Wu X, Lü G, Gao H (2023) Drought stress tolerance in vegetables: the functional role of structural features, key gene pathways, and exogenous hormones. Int J Mol Sci 24(18):13876. https://doi.org/10.3390/ijms241813876
2. Abramova A, Vereshchagin M, Kulkov L, Kreslavski VD, Kuznetsov VV, Pashkovskiy P (2023) Potential role of phytochromes A and B and cryptochrome 1 in the adaptation of Solanum lycopersicum to UV-B radiation. Int J Mol Sci 24(17):13142. https://doi.org/10.3390/ijms241713142
3. Ahammed GJ, Gantait S, Mitra M, Yang Y, Li X (2020) Role of ethylene crosstalk in seed germination and early seedling development: a review. Plant Physiol Biochem 151:124–131. https://doi.org/10.1016/j.plaphy.2020.03.016
4. Ahrens A, Llewellyn D, Zheng Y (2023) Is twelve hours the optimum photoperiod for promoting flowering in indoor-grown cultivars of Cannabis sativa? Plants 12(14):2605. https://doi.org/10.3390/plants12142605

5. Al-Yasi H, Attia H, Alamer K, Hassan F, Ali E, Elshazly S, Hessini K et al (2020) Impact of drought on growth, photosynthesis, osmotic adjustment, and cell wall elasticity in Damask rose. Plant Physiol Biochem 150:133–139. https://doi.org/10.1016/j.plaphy.2020.02.038
6. Ameur F, Amichi H, Leauthaud C (2020) Agroecology in North African irrigated plains? Mapping promising practices and characterizing farmers' underlying logics. Reg Environ Change 20(4):133. https://doi.org/10.1007/s10113-020-01719-1
7. Banaś AK, Leja K, Zgłobicki P, Jedynak P, Kowalska E, Strzałka W, Myśliwa-Kurdziel B et al (2024) De-etiolation is almost color blind: the study of photosynthesis awakening under blue and red light. Plant Cell Physiol 65(12):1993–2017. https://doi.org/10.1093/pcp/pcae119
8. Baslam M, Mitsui T, Hodges M, Priesack E, Herritt MT, Aranjuelo I, Sanz-Sáez Á (2020) Photosynthesis in a changing global climate: scaling up and scaling down in crops. Front Plant Sci 11:882. https://doi.org/10.3389/fpls.2020.00882
9. Bawa G, Feng L, Chen G, Chen H, Hu Y, Pu T, Wang X et al (2020) Gibberellins and auxin regulate soybean hypocotyl elongation under low light and high-temperature interaction. Physiol Plant 170(3):345–356. https://doi.org/10.1111/ppl.13158
10. Bhatla SC, Lal MA (2023). Light perception and transduction. In: Plant physiology, development and metabolism. Springer Nature Singapore, Singapore, pp 363–390. https://doi.org/10.1007/978-981-99-5736-1_13
11. Bedeke SB (2023) Climate change vulnerability and adaptation of crop producers in sub-Saharan Africa: a review on concepts, approaches and methods. Environ Dev Sustain 25(2):1017–1051. https://doi.org/10.1007/s10668-022-02118-8
12. Brini F, Mseddi K, Brestic M, Landi M (2022) Hormone-mediated plant responses to light quality and quantity. Environ Exp Bot 202:105026. https://doi.org/10.1016/j.envexpbot.2022.105026
13. Buttò V, Rozenberg P, Deslauriers A, Rossi S, Morin H (2021) Environmental and developmental factors driving xylem anatomy and micro-density in black spruce. New Phytol 230(3):957–971. https://doi.org/10.1111/nph.17223
14. Cai Z, Zhang Y, Tang W, Chen X, Lin C, Liu Y, Duan Y et al (2022) LUX ARRHYTHMO interacts with ELF3a and ELF4a to coordinate vegetative growth and photoperiodic flowering in rice. Front Plant Sci 13:853042. https://doi.org/10.3389/fpls.2022.853042
15. Castro-Camba R, Sánchez C, Vidal N, Vielba JM (2022) Plant development and crop yield: the role of gibberellins. Plants 11(19):2650. https://doi.org/10.3390/plants11192650
16. Dang X, Zhang Y, Li Y, Chen S, Liu E, Fang B, Hong D et al (2022) SYL3-k increases style length and yield of F1 seeds via enhancement of endogenous GA4 content in Oryza sativa L. pistils. Theor Appl Genet 135(1):321–336. https://doi.org/10.1007/s00122-021-03968-y
17. Demski C, Capstick S, Pidgeon N, Sposato RG, Spence A (2017) Experience of extreme weather affects climate change mitigation and adaptation responses. Clim Change 140:149–164. https://doi.org/10.1007/s10584-016-1837-4
18. Dubo T, Palomo I, Camacho LL, Locatelli B, Cugniet A, Racinais N, Lavorel S (2023) Nature-based solutions for climate change adaptation are not located where they are most needed across the Alps. Reg Environ Change 23(1):12. https://doi.org/10.1007/s10113-022-01998-w
19. Dwivedi SL, Quiroz LF, Spillane C, Wu R, Mattoo AK, Ortiz R (2024) Unlocking allelic variation in circadian clock genes to develop environmentally robust and productive crops. Planta 259(4):72. https://doi.org/10.1007/s00425-023-04324-8
20. Ettinger AK, Buonaiuto DM, Chamberlain CJ, Morales-Castilla I, Wolkovich EM (2021) Spatial and temporal shifts in photoperiod with climate change. New Phytol 230(2):462–474. https://doi.org/10.1111/nph.17172
21. Fantini E, Facella P (2020) Cryptochromes in the field: how blue light influences crop development. Physiol Plant 169(3):336–346. https://doi.org/10.1111/ppl.13088
22. Falcón J, Torriglia A, Attia D, Viénot F, Gronfier C, Behar-Cohen F, Hicks D et al (2020) Exposure to artificial light at night and the consequences for flora, fauna, and ecosystems. Front Neurosci 14:602796. https://doi.org/10.3389/fnins.2020.602796
23. Ferreyra MLF, Serra P, Casati P (2021) Recent advances on the roles of flavonoids as plant protective molecules after UV and high light exposure. Physiol Plant 173(3):736–749. https://doi.org/10.1111/ppl.13543

24. Fraikin GY, Belenikina NS, Rubin AB (2023) Molecular bases of signaling processes regulated by cryptochrome sensory photoreceptors in plants. Biochem Mosc 88(6):770–782. https://doi.org/10.1134/S0006297923060056
25. Freytes SN, Canelo M, Cerdán PD (2021) Regulation of flowering time: when and where? Curr Opin Plant Biol 63:102049. https://doi.org/10.1016/j.pbi.2021.102049
26. Gupta N, Nath U (2020) Integration of light and hormone response during seedling establishment. J Plant Biochem Biotechnol 29(4):652–664. https://doi.org/10.1007/s13562-020-00628-y
27. Han H, Adamowski M, Qi L, Alotaibi SS, Friml J (2021) PIN-mediated polar auxin transport regulations in plant tropic responses. New Phytol 232(2):510–522. https://doi.org/10.1111/nph.17617
28. Hessel V, Liang S, Tran NN, Escribà-Gelonch M, Zeckovic O, Knowling M, Burgess A et al (2022) Eustress in space: opportunities for plant stressors beyond the earth ecosystem. Front Astron Space Sci 9:841211. https://doi.org/10.3389/fspas.2022.841211
29. Hu W, Figueroa-Balderas R, Chi-Ham C, Lagarias JC (2020) Regulation of monocot and dicot plant development with constitutively active alleles of phytochrome B. Plant Direct 4(4):e00210. https://doi.org/10.1002/pld3.210
30. Huq E, Lin C, Quail PH (2024) Light signaling in plants—a selective history. Plant Physiol 195(1):213–231. https://doi.org/10.1093/plphys/kiae110
31. Izzo LG, Mele BH, Vitale L, Vitale E, Arena C (2020) The role of monochromatic red and blue light in tomato early photomorphogenesis and photosynthetic traits. Environ Exp Bot 179:104195. https://doi.org/10.1016/j.envexpbot.2020.104195
32. Janeeshma E, Johnson R, Amritha MS, Noble L, Aswathi KR, Telesiński A, Puthur JT et al (2022) Modulations in chlorophyll a fluorescence based on intensity and spectral variations of light. Int J Mol Sci 23(10):5599. https://doi.org/10.3390/ijms23105599
33. Jarad M, Antoniou-Kourounioti R, Hepworth J, Qüesta JI (2020) Unique and contrasting effects of light and temperature cues on plant transcriptional programs. Transcription 11(3–4):134–159. https://doi.org/10.1080/21541264.2020.1820299
34. Jing Y, Lin R (2020) Transcriptional regulatory network of the light signaling pathways. New Phytol 227(3):683–697. https://doi.org/10.1111/nph.16602
35. Kaiser E, Kusuma P, Vialet-Chabrand S, Folta K, Liu Y, Poorter H, Marcelis LF et al (2024) Vertical farming goes dynamic: optimizing resource use efficiency, product quality, and energy costs. Front Sci 2:1411259. https://doi.org/10.3389/fsci.2024.1411259
36. Kang SG, Lee KE, Singh M, Kumar P, Matin MN (2021) Rice lesion mimic mutants (LMM): the current understanding of genetic mutations in the failure of ROS scavenging during lesion formation. Plants 10(8):1598. https://doi.org/10.3390/plants10081598
37. Kanojia A, Bhola D, Mudgil Y (2023) Light signaling as cellular integrator of multiple environmental cues in plants. Physiol Mol Biol Plants 29(10):1485–1503. https://doi.org/10.1007/s12298-023-01364-y
38. Karki N, Vergish S, Zoltowski BD (2021) Cryptochromes: Photochemical and structural insight into magnetoreception. Protein Sci 30(8):1521–1534. https://doi.org/10.1002/pro.4124
39. Kharshiing EV, Mawphlang OIL, Lama V, Bhattacharjee R, Sahoo L (2022) Manipulation of light environment for optimising photoreceptor activity enhancing plant traits of agronomic and horticultural importance in crops. J Hortic Sci Biotechnol 97(5):535–551. https://doi.org/10.1080/14620316.2022.2056089
40. Khudyakova AY, Kosobryukhov AA, Pashkovskiy PP, Kreslavski VD (2024) Cryptochromes and their role in the process of plant adaptation. Russ J Plant Physiol 71(1):42. https://doi.org/10.1134/S102144372360366X
41. Kishchenko O, Zhou Y, Jatayev S, Shavrukov Y, Borisjuk N (2020) Gene editing applications to modulate crop flowering time and seed dormancy. Abiotech 1(4):233–245. https://doi.org/10.1007/s42994-020-00032-z
42. Kutschera U (2015) Basic versus applied research: Julius Sachs (1832–1897) and the experimental physiology of plants. Plant Signal Behav 10(9):e1062958. https://doi.org/10.1002/ajb2.1078

43. Laad P, Patel P, Guruprasad KN (2023) UVR8 signaling, mechanism, and integration with other pathways. In: UV-B radiation and crop growth. Springer Nature Singapore, Singapore, pp 193–221. https://doi.org/10.1007/978-981-19-3620-3_10
44. Landi M, Zivcak M, Sytar O, Brestic M, Allakhverdiev SI (2020) Plasticity of photosynthetic processes and the accumulation of secondary metabolites in plants in response to monochromatic light environments: a review. Biochimica et Biophysica Acta (BBA)-Bioenergetics 1861(2):148131. https://doi.org/10.1016/j.bbabio.2019.148131
45. Lebedeva MA, Dodueva IE, Gancheva MS, Tvorogova VE, Kuznetsova KA, Lutova LA (2020) The evolutionary aspects of flowering control: florigens and anti-florigens. Russ J Genet 56:1323–1344. https://doi.org/10.1134/S102279542011006X
46. Lee Z, Kim S, Choi SJ, Joung E, Kwon M, Park HJ, Shim JS (2023) Regulation of flowering time by environmental factors in plants. Plants 12(21):3680. https://doi.org/10.3390/plants12213680
47. Lee Z, Lim JA, Harikrishna JA, Islam T, Abd Rahim MH, Yaacob JS (2024) Regulation of plant responses to temperature stress: a key factor in food security and for mitigating effects of climate change. Int J Plant Prod 18(2):141–159. https://doi.org/10.1007/s42106-024-00282-7
48. Lei H, Su S, Ma L, Wen Y, Wang X (2017) Molecular cloning and functional characterization of CoFT1, a homolog of FLOWERING LOCUS T (FT) from Camellia oleifera. Gene 626:215–226. https://doi.org/10.1016/j.gene.2017.05.044
49. Li Y, Xin G, Shi Q, Yang F, Wei M (2023) Response of photomorphogenesis and photosynthetic properties of sweet pepper seedlings exposed to mixed red and blue light. Front Plant Sci 13:984051. https://doi.org/10.3389/fpls.2022.984051
50. Lian X, Jiao L, Zhong J, Jia Q, Liu J, Liu Z (2021) Artificial light pollution inhibits plant phenology advance induced by climate warming. Environ Pollut 291:118110. https://doi.org/10.1016/j.envpol.2021.118110
51. Liebers M, Pfannschmidt T (2024) New horizons in light control of plant photomorphogenesis and development. Front Photobiol 1:1346705. https://doi.org/10.3389/fphbi.2023.1346705
52. Lymperopoulos P, Msanne J, Rabara R (2018) Phytochrome and phytohormones: working in tandem for plant growth and development. Front Plant Sci 9:1037. https://doi.org/10.3389/fpls.2018.01037
53. Modarelli GC, Arena C, Pesce G, Dell'Aversana E, Fusco GM, Carillo P, Paradiso R et al (2020) The role of light quality of photoperiodic lighting on photosynthesis, flowering and metabolic profiling in Ranunculus asiaticus L. Physiol Plant 170(2):187–201. https://doi.org/10.1111/ppl.13122
54. Morales F, Ancín M, Fakhet D, González-Torralba J, Gámez AL, Seminario A, Aranjuelo I et al (2020) Photosynthetic metabolism under stressful growth conditions as a bases for crop breeding and yield improvement. Plants 9(1):88. https://doi.org/10.3390/plants9010088
55. Nowicka B, Kruk J (2016) Powered by light: phototrophy and photosynthesis in prokaryotes and its evolution. Microbiol Res 186:99–118. https://doi.org/10.1016/j.micres.2016.04.001
56. Oakenfull RJ, Davis SJ (2017) Shining a light on the Arabidopsis circadian clock. Plant, Cell Environ 40(11):2571–2585. https://doi.org/10.1111/pce.13033v
57. Oide M, Hikima T, Oroguchi T, Kato T, Yamaguchi Y, Yoshihara S, Okajima K et al (2020) Molecular shape under far-red light and red light-induced association of Arabidopsis phytochrome B. FEBS J 287(8):1612–1625. https://doi.org/10.1111/febs.15095
58. Osnato M, Cota I, Nebhnani P, Cereijo U, Pelaz S (2022) Photoperiod control of plant growth: flowering time genes beyond flowering. Front Plant Sci 12:805635. https://doi.org/10.3389/fpls.2021.805635
59. Paradiso R, Proietti S (2022) Light-quality manipulation to control plant growth and photomorphogenesis in greenhouse horticulture: The state of the art and the opportunities of modern LED systems. J Plant Growth Regul 41(2):742–780. https://doi.org/10.1007/s00344-021-10337-y
60. Pattaro MC, Falcioni R, Moriwaki T, Alves DC, Antunes WC (2024) Blue light strongly promotes de-etiolation over green, moderate over red, but have limited action over far-red lights in lettuce plants. Sci Hortic 328:112863. https://doi.org/10.1016/j.scienta.2024.112863

61. Pennisi G, Pistillo A, Orsini F, Cellini A, Spinelli F, Nicola S, Marcelis LF et al (2020) Optimal light intensity for sustainable water and energy use in indoor cultivation of lettuce and basil under red and blue LEDs. Sci Hortic 272:109508. https://doi.org/10.1016/j.scienta.2020.109508
62. Percio F, Rubio L, Amorim-Silva V, Botella MA (2025) Crucial roles of brassinosteroids in cell wall composition and structure across species: new insights and biotechnological applications. Plant, Cell Environ 48(3):1751–1767. https://doi.org/10.1111/pce.15258
63. Pierik R, Ballaré CL (2021) Control of plant growth and defense by photoreceptors: from mechanisms to opportunities in agriculture. Mol Plant 14(1):61–76. https://doi.org/10.1016/j.molp.2020.11.021
64. Proietti S, Scariot V, De Pascale S, Paradiso R (2022) Flowering mechanisms and environmental stimuli for flower transition: bases for production scheduling in greenhouse floriculture. Plants 11(3):432. https://doi.org/10.3390/plants11030432
65. Raspor M, Motyka V, Kaleri AR, Ninković S, Tubić L, Cingel A, Ćosić T (2021) Integrating the roles for cytokinin and auxin in de novo shoot organogenesis: from hormone uptake to signaling outputs. Int J Mol Sci 22(16):8554. https://doi.org/10.3390/ijms22168554
66. Razzaq K, Du J (2024) Phytohormonal regulation of plant development in response to fluctuating light conditions. J Plant Growth Regul 1–34. https://doi.org/10.1007/s00344-024-11568-5
67. Rehman M, Pan J, Mubeen S, Ma W, Luo D, Cao S, Chen P et al (2024) Morpho-physio-biochemical, molecular, and phytoremedial responses of plants to red, blue, and green light: a review. Environ Sci Pollut Res 31(14):20772–20791. https://doi.org/10.1007/s11356-024-32532-6
68. Sanchez SE, Rugnone ML, Kay SA (2020) Light perception: a matter of time. Mol Plant 13(3):363–385. https://doi.org/10.1016/j.molp.2020.02.006
69. Santin M, Simoni S, Vangelisti A, Giordani T, Cavallini A, Mannucci A, Castagna A et al (2023) Transcriptomic analysis on the peel of UV-B-exposed peach fruit reveals an upregulation of phenolic-and UVR8-related pathways. Plants 12(9):1818. https://doi.org/10.3390/plants12091818
70. Sarwar R, Zhu KM, Jiang T, Ding P, Gao Y, Tan XL (2023) DELLAs directed gibberellins responses orchestrate crop development: a brief review. Crop Sci 63(1):1–28. https://doi.org/10.1002/csc2.20825
71. Shah SH, Islam S, Mohammad F, Siddiqui MH (2023) Gibberellic acid: a versatile regulator of plant growth, development and stress responses. J Plant Growth Regul 42(12):7352–7373. https://doi.org/10.1007/s00344-023-11035-7
72. Sharma A, Samtani H, Sahu K, Sharma AK, Khurana JP, Khurana P (2023) Functions of phytochrome-interacting factors (PIFs) in the regulation of plant growth and development: a comprehensive review. Int J Biol Macromol 244:125234. https://doi.org/10.1016/j.ijbiomac.2023.125234
73. Singh P, Arif Y, Siddiqui H, Upadhyaya CP, Pichtel J, Hayat S (2023) Critical factors responsible for potato tuberization. Bot Rev 89(4):421–437. https://doi.org/10.1007/s12229-023-09289-7
74. Stanton D, Justin HS, Reitzel AM (2022) Step in time: conservation of circadian clock genes in animal evolution. Integr Comp Biol 62(6):1503–1518. https://doi.org/10.1093/icb/icac140
75. Su C, Wang Y, Yu Y, He Y, Wang L (2021) Coordinative regulation of plants growth and development by light and circadian clock. Abiotech 2(2):176–189. https://doi.org/10.1007/s42994-021-00041-6
76. Sun C, He C, Zhong C, Liu S, Liu H, Luo X, Deng X et al (2022) Bifunctional regulators of photoperiodic flowering in short day plant rice. Front Plant Sci 13:1044790. https://doi.org/10.3389/fpls.2022.1044790
77. Takagi H, Hempton AK, Imaizumi T (2023) Photoperiodic flowering in arabidopsis: multilayered regulatory mechanisms of CONSTANS and the florigen FLOWERING LOCUS T. Plant Commun 4(3). https://doi.org/10.1016/j.xplc.2023.100552
78. Tan H, Qi X, Li Y, Wang X, Zhou J, Liu X, Xiang X et al (2020) Light induces gene expression to enhance the synthesis of storage reserves in Brassica napus L. embryos. Plant Mol Biol 103:457–471. https://doi.org/10.1007/s11103-020-01003-0

79. Tan P, He L, Huang Y, Zhou Y (2022) Optophysiology: Illuminating cell physiology with optogenetics. Physiol Rev 102(3):1263–1325. https://doi.org/10.1152/physrev.00021.2021
80. Thakur T, Sinha K, Kaur T, Kapoor R, Kumar G, Bhunia RK, Salvi P (2022) Efficient genetic transformation of rice for CRISPR/Cas9 mediated genome-editing and stable overexpression studies: a case study on rice lipase 1 and galactinol synthase encoding genes. Agronomy 12(1):179. https://doi.org/10.3390/agronomy12010179
81. Tripathi S, Hoang QT, Han YJ, Kim JI (2019) Regulation of photomorphogenic development by plant phytochromes. Int J Mol Sci 20(24):6165. https://doi.org/10.3390/ijms20246165
82. Thies M, Ferrarezi RS, Realff M (2024) Combined effect of light and temperature on wheat and rice growth: A case study in controlled environment agriculture. Agronomy 14(8):1641. https://doi.org/10.3390/agronomy14081641
83. Vaishnav D, Chowdhury P (2023) Types and function of phytohormone and their role in stress. In: Plant abiotic stress responses and tolerance mechanisms. IntechOpen.
84. van Zanten M, Ai H, Quint M (2021) Plant thermotropism: an underexplored thermal engagement and avoidance strategy. J Exp Bot 72(21):7414–7420. https://doi.org/10.1093/jxb/erab209
85. Verheul MJ, Maessen HF, Paponov M, Panosyan A, Kechasov D, Naseer M, Paponov IA (2022) Artificial top-light is more efficient for tomato production than inter-light. Sci Hortic 291:110537. https://doi.org/10.1016/j.scienta.2021.110537
86. Wang P, Abid MA, Qanmber G, Askari M, Zhou L, Song Y, Zhang R et al (2022) Photomorphogenesis in plants: the central role of phytochrome interacting factors (PIFs). Environ Exp Bot 194:104704. https://doi.org/10.1016/j.envexpbot.2021.104704
87. Wang S, Liu X, Liu X, Xue J, Ren X, Zhai Y, Zhang X (2022) The red/blue light ratios from light-emitting diodes affect growth and flower quality of Hippeastrum hybridum 'Red Lion.' Front Plant Sci 13:1048770. https://doi.org/10.3389/fpls.2022.1048770
88. Xiong B, Qiu X, Huang S, Wang X, Zhang X, Dong T, Wang Z et al (2019) Physiological and transcriptome analyses of photosynthesis and chlorophyll metabolism in variegated Citrus (Shiranuhi and Huangguogan) seedlings. Sci Rep 9(1):15670. https://doi.org/10.1038/s41598-019-52276-5
89. Yadav A, Singh D, Lingwan M, Yadukrishnan P, Masakapalli SK, Datta S (2020) Light signaling and UV-B-mediated plant growth regulation. J Integr Plant Biol 62(9):1270–1292. https://doi.org/10.1111/jipb.12932
90. Yadukrishnan P, Singh D, Ravindran N, Kushwaha AK, Job N, Rahul PV, Datta S et al (2021) Hormones and light-regulated seedling development. In: Hormones and plant response. Springer International Publishing, Cham, pp 91–116. https://doi.org/10.1007/978-3-030-77477-6_4
91. Yang Z, Liu B, Su J, Liao J, Lin C, Oka Y (2017) Cryptochromes orchestrate transcription regulation of diverse blue light responses in plants. Photochem Photobiol 93(1):112–127. https://doi.org/10.1111/php.12663
92. Yang X, Wang S, Liu W, Huang S, Xie Y, Meng X, Yu J et al (2025) Different spatial configurations of LED light sources enhance growth in tomato seedlings by influencing photosynthesis, CO_2 assimilation, and endogenous hormones. Plants 14(9):1369. https://doi.org/10.3390/plants14091369
93. Zhang Q, Li X, Cai S, Li J, Wang J, Li Y, Dai S (2024) Flowering characteristics and expression patterns of key genes in response to photoperiod in different chrysanthemum varieties. Horticulturae 11(1):5. https://doi.org/10.3390/horticulturae11010005
94. Zong W, Guo X, Zhang K, Chen L, Liu YG, Guo J (2024) Photoperiod and temperature synergistically regulate heading date and regional adaptation in rice. J Exp Bot 75(13):3762–3777. https://doi.org/10.1093/jxb/erae209

Chapter 8
Cell Division in Plants is Influenced by Diurnal Rhythm

Prasann Kumar, Shipa Rani Dey, Joginder Singh, and Neha Gogia

1 Introduction

However, cellular division is a fundamental biological process at the heart of all organisms' growth and development, including plants. It reflects the mechanisms by which a mother cell undergoes division, producing two daughter cells, and whereby its genetic material is accurately transferred to the subsequent cell generations. In plants, it is critical for the growth of individual cells and the new tissues, organs, and, ultimately, the whole plant. Cell division can be divided into two major processes: mitosis, where the genetic material of the cell is separated, and cytokinesis, where the cytoplasm is divided, forming two distinct cells [1, 18]. Cell division in plants is crucial for maintaining the balance of plant tissues. This is paramount during developmental phases like seed germination, root and shoot lengthening, flower and fruit formation, and the restoration of damaged plant structures. In addition, as plants become more complex, cell division also accounts for the number of cells in a plant and provides the building blocks for total plant biomass and plant structure.

The plant cell cycle is the set of actions that must happen in an ordered manner to divide a plant cell correctly and at the proper moment, and as such, it is among the most important features of plant cell division. Internal and external cues, including

P. Kumar (✉)
Department of Agronomy, School of Agriculture, Lovely Professional University, Phagwara, Punjab, India
e-mail: prasann0659@gmail.com

S. R. Dey
Department of Agronomy, Faculty of Agricultural Sciences, GLA University, Chaumuhan, Uttar Pradesh, India

J. Singh
Department of Botany, Nagaland University, Nagaland, India

N. Gogia
New Haven, CT, USA

A. Kothari Chhajer and V. Singh Gour (eds.), *Introduction to Rhythms in Plants*, Studies in Rhythm Engineering, https://doi.org/10.1007/978-981-95-3592-7_8

environmental signals like light, temperature, and humidity, drive the regulation. Of the environmental factors, light is essential in regulating cellular division in plants. Plants have developed complex systems to coordinate internal cellular processes with the external environment to optimize growth. Multiple mechanisms, including diurnal rhythms that correlate with regulating cell division, accomplish such synchronization [9, 20].

Diurnal rhythms are biological processes that change in a cycle of around 24 h, depending upon external environmental processes, primarily light and darkness. Most living organisms (including plants), though, experience these rhythms as they control a number of their physiological processes, including photosynthesis, respiration, and growth. In plants, diurnal rhythms are mainly regulated by an internal circadian clock, which allows plants to anticipate variations in the day-night cycles. This internal circadian clock organizes a plant's biological functions well ahead of time, helping it to maximize its metabolic activities according to the light conditions it has to respond to outside. Plants also use photosynthesis to convert light into chemical energy during the day, respiration processes, and growth at night. Circadian rhythm ensures that these processes are carried out promptly for maximum productivity [5, 11].

This leads to the undeniable need for daily cycles regulating biological processes, especially in the plant kingdom. These cycles assist plants in adapting to daily environmental variability, such as differences in light intensity, temperature, and humidity. Diurnal rhythms dictate the timing of central events like stomata closure and opening, biosynthesis of important plant hormones, and gene expression patterns. The plant has adapted these life processes (photosynthesis, respiration, etc.) to the environmental cues received by the day-night cycle, and the synchronization of these processes is the key to the planet's survival by ensuring life processes will not be disrupted [13, 17]. The circadian clock, for example, regulates when enzymes that participate in photosynthesis are synthesized, producing them when light is present and active when needed. It also orchestrates phenomena such as leaf and flower movement, which can depend on the brightness level, so that plants can capture as much sunlight as possible.

This link between organ development and circadian rhythms is essential for the proper growth and function of plants. The plant's circadian clock gives rise to a rhythmic division of cells, with the timing of cell division being controlled at specific times during the light and dark periods. This temporal control over cell division is necessary to coordinate the plant's growth in relation to its available resources (primarily light and nutrients), which vary throughout the day. In the daytime, a high amount of energy is provided by photosynthesis, and the plant is most likely to build new cells, especially in the meristematic tissues like root and shoot tips, elongating the plant. Since the energy from photosynthesis is not available at night, the plant could reduce cell division and focus on nutrient uptake and growth [4, 25].

In many plants, cell division peaks during the daytime when light conditions are optimal for photosynthesis. Rapid cell growth is significant for tissues like meristems, where cells divide to create populations of cells that could later differentiate into specialized tissues. This rhythm helps ensure that energy harvested during daytime is metabolized through cell division and growth at peak efficiency. By coordinating these activities to a diurnal rhythm, diurnal rhythms prevent the plant from wasting energy through unnecessary cell division when energy is not abundant, such as during the night or under unfavorable environmental conditions. Synchronizing these processes with diurnal rhythms allows the plant to maximize its growth potential by aligning energy production, storage, and utilization [3, 10].

Diurnal rhythms also affect the expression of major cell cycle regulators, so cell division corresponds with the entire organism's physiological state. The cell cycle gene expression timing correlates with the plant's circadian clock. This allows the plant to time the activation of the genes responsible for cell division with both its internal metabolic state and external environmental conditions. This way, the plant optimizes its growth potential while reducing the likelihood of wasting energy or damaging tissue. For instance, plant hormones such as auxins, cytokinin, and gibberellins, which all regulate the division of cells, are also governed by circadian rhythms, meaning that they are present in the correct location and in the right amounts to stimulate division when the plant requires growth.

Moreover, aligning cell division with diurnal rhythms prevents unnecessary energy consumption and stress. A wide range of abiotic stresses, such as changes in temperature, water availability, and nutrient supply, constantly challenge plants. Because environmental conditions can vary significantly throughout the day, this coordination of cell division with the plant's circadian rhythm allows the plant to better adjust to these changes, as cell division will occur when resources are most plentiful and growth is most adequate [14, 16]. The competing influence of water availability, which slows or stops cell division to conserve resources, can divert energy to maintain cell integrity and avoid unnecessary metabolism expenditure.

Places can optimize their available resources by coordinating cell division with the day-night cycle, balancing energy production and storage, and adapting to environmental conditions. Such synchronization assures that plants grow and repair damaged tissues as needed and form new organs appropriately. Diurnal rhythms control key functions in plant cell life, including cell division; understanding these mechanisms may lead to solutions to the ongoing challenges of improving productive growth and stress resistance in plants, which are critical to sustainable agriculture under climate change, among other pressing environmental challenges.

2 Cell Cycle Phases in Plants

What is the cell cycle? The plant cell cycle is a complex and tightly regulated series of events resulting in the cell's division, growth, and development. This separates the parent cell into two daughter cells; both must get a complete set of chromosomes and enough raw materials to survive and multiply. The cell cycle is a critical regulator of organ growth, tissue regeneration, and response to plant environmental cues. Cell Cycle. The cell cycle is generally divided into two broad phases: interphase and mitotic phase (M phase). Interphase consists of the G1, S, and G2 phases, where the cell grows, replicates its DNA, and prepares for division. This contrasts with the M phase, in which the cell divides into two daughter cells with similar genetic material. These phases are tightly regulated to guarantee normal plant development and proper functioning; aberrations in the cell cycle can cause developmental defects, compromised growth, or, ultimately, cell death [8, 21].

The G1 phase (Gap 1) is the first phase of interphase and is characterized by cell growth and preparation for DNA replication. The cell completes various metabolic functions in this stage, from synthesizing proteins and lipids to collecting the necessary components for DNA replication. The G1 phase is essential in sensing environmental information, such as the availability of light, water, and nutrients in plant cells. It influences whether a plant organ will continue its cell cycle or withdraw, entering quiescence. Say, for instance, that the plant faces adverse conditions like drought or nutrient deficiency. If that is the case, the cell could enter a quiescent state called G0 that pauses its journey in the cell cycle until conditions become more favorable. This G1-type ability to pause the cell cycle allows plants to conserve energy and resources and prevent division when environmental parameters are non-permissive to growth. After amassing enough resources, the cell enters the S phase (synthesis phase), which begins dividing [15, 23].

During the S phase, DNA replication takes place. This is a stage where the cell replicates its complete genome so that each daughter cell has an exact copy of the genome. As DNA replication must be accurate to avoid mutagenesis, many checkpoints exist to identify and repair errors that arise during DNA replication. Despite how plants replicate DNA, it does not happen randomly and is tightly regulated by proteins and regulatory pathways throughout the cell cycle. Following DNA replication, the cell enters the G2 phase (Gap 2), during which the cell grows and starts to prepare for mitosis. In the G2 phase, the cell creates and gathers proteins and organelles that it will use during cell division and assesses whether any mistakes have been made in the replicated DNA. If any errors are noticed, the cell will try to fix them, and then the M phase will follow. Besides, the G2 phase is critical for energy and material accumulation before mitosis [7, 24].

The M phase, or mitotic phase, is the final stage of the cell cycle, during which a cell divides its nucleus and cytoplasm to form two genetically identical daughter cells. This process includes two main events: mitosis (division of the nucleus) and cytokinesis (division of the cytoplasm).

Mitosis is a highly organized process that occurs in four stages: prophase, metaphase, anaphase, and telophase [8, 10, 24].

- In prophase, the loosely packed chromatin condenses into visible chromosomes, and the nuclear envelope begins to break down.
- During metaphase, the chromosomes line up at the center of the cell, called the equator, ready to be separated.
- In anaphase, the sister chromatids (identical copies of each chromosome) are pulled apart to opposite ends (poles) of the cell by thread-like structures called spindle fibers.
- In telophase, the separated chromatids reach the poles, and new nuclear envelopes form around them, creating two distinct nuclei.

Finally, cytokinesis takes place, where the cytoplasm divides, resulting in two separate daughter cells. This entire process must be tightly controlled to ensure that each new cell receives an exact copy of the genetic material and sufficient cellular resources [2, 12]. Proper regulation of the M phase is essential for healthy growth, development, and tissue repair in plants and animals.

Cyclins and cyclin-dependent kinases (CDKs) act as molecular regulators that tightly control the progression of the cell cycle in plants. These regulators synchronize the stepwise progression through different cell cycle phases to execute each phase in the correct sequence and promptly. Cyclins are a group of proteins whose expression varies during the cell cycle by binding to and activating CDKs, enzymes that phosphorylate target proteins, to modulate cell cycle progression. Cyclins and CDKs are regulated by multiple mechanisms, including cyclin synthesis and degradation, as well as the binding of CDK inhibitors, which inhibit them when their activity is not needed. Different types of cyclins are distinguished by the phases they control, namely G1 cyclins, S cyclins, and M cyclins, which prevent the progression of the G1 phase, S phase, and M phase, respectively [4, 16, 23]. Cyclins promote CDKs in the G1 phase that push the cell to duplicate its DNA, and cyclins activate CDKs in the M phase, which is responsible for managing the processes of mitosis. Cyclin-associated Cyclin-dependent kinase (CDK) complexes are carefully regulated to keep the order of cell cycle events sequential and unique for each phase. The degradation of cyclins at the end of each phase ensures that the cell does not go to the next phase before it is ready and that the cell cycle phases are correctly coordinated [6, 20, 22].

Apart from cyclins and CDKs, several other major regulatory proteins participate in modulating the cell cycle in plants. These include cyclin-dependent kinase inhibitors (CKIs), which inhibit CDKs from activating, and checkpoints that inspect the integrity of the cell cycle to allow or stop progression if issues are detected. In the S phase, if DNA damage occurs, checkpoint proteins can halt the cycle to permit DNA repair. The cell could go through apoptosis, a programmed cell death, if the damage is too bad to prevent the proliferation of damaged cells. In plants, these regulatory mechanisms coordinate cell division in a controlled fashion, which is essential for growth, development, and health.

Regulated cell cycle development is critical for plant health and vigor. Disruption of the cell cycle regulatory framework may lead to developmental defects, dysfunctional tissue growth, or compromised plant survival. Some potential implications include stunting and delayed maturation; there is likely also impaired responsiveness to environmental pressures if cells fail to divide correctly. In contrast, uncontrolled cell division, seen in tumors or cancer, can disturb tissues' normal growth and organization. These aberrant cell divisions can lead to plant calluses or galls, disrupting plant function and developmental processes [2, 16, 23]. Disrupted cell cycle regulation can also affect the plant's ability to respond to environmental signals, including light and water availability. It could make it harder for the plant to adapt to shifting conditions.

Plant cell cycle regulation has been extensively studied using model organisms, including *Arabidopsis thaliana*. *Arabidopsis* is a plant species recognized as a model organism for plant biology that can be grown quickly. It has a relatively more straightforward genome and well-characterized genetics. *Arabidopsis*, as a model organism, is genetically manipulatable, enabling scientists to examine the role of individual genes in regulating cell cycle progression by overexpressing or knocking down specific genes involved in the cell cycle to gain insight into their role in plant growth and development. The initial discoveries in *Arabidopsis* have identified cyclins and cyclin-dependent kinases (CDKs) that regulate the passage through specific phases of the cell cycle and checkpoint proteins that monitor the integrity of DNA replication, for example. These discoveries have significant implications for plant growth and development, as well as for crop productivity and stress tolerance [10, 21, 24].

These replication-independent epigenetic mechanisms maintain gene expression states as plant cells progress through the tightly regulated, orderly transition through the G1, S, G2, and M phases we now refer to as the plant cell cycle. Cyclins and CDKs, important cell cycle regulators, aid in timing events, ensuring cells divide on time and under the right circumstances. Cell cycle regulation plays a key role in development, growth, and survival in the plant life cycle. Thus, inappropriate cell cycle progression is often detrimental to plant health and growth, giving rise to developmental defects, stunted growth, or reduced survival [7, 13, 14]. *Arabidopsis* and other model organisms have played a pivotal role in advancing our understanding of how plants control their cell cycles, yielding critical molecular insights into regulating cell division and plant growth. Gaining insights into these processes enhances our understanding of the biological basis for plant growth and development. It strives for improved crop productivity and increased plant resilience to environmental adversities (Table 1).

3 Circadian Clock Control of Cell Division

The circadian clock is an indispensable biological timer that regulates different physiological processes, from plants to fungi and humans. This internal clock coordinates various methods, like cell division, to synchronize these activities with the optimal time of day. The circadian clock plays a vital role in synchronizing cellular events, including the cell cycle, at the cell level within an organism to respond to specific environmental inputs such as light, temperature, or nutrients (based on the availability) in plants. The plant circadian clock consists of a network of genes and proteins that regulate its function through feedback loops. Circadian rhythms are encoded by transcriptional activators in the plant circadian clock, like Circadian Clock Associated 1 (CCA1), Late Elongated Hypocotyl (LHY), and Timeless and Clock 1 (TOC1). These factors regulate downstream genes required for normal cellular processes, including division. The link between circadian rhythms and the cell cycle governs the timing of plant cell division, fine-tuning plant growth and development, and enabling the plant to adapt to environmental changes [10, 22–24].

The circadian clock is a complex molecular circuit comprised of multiple interlocking feedback loops, which generate rhythmic oscillations of gene expression daily. Two of the most vital components in regulating the plant circadian clock are the

Table 1 Cell cycle phases and key processes for gene expression

Cell cycle phase	Key processes	Regulatory proteins	Gene expression	Time of occurrence	Checkpoint mechanisms	Significance in plants	Experimental methods
G1 Phase	Cell growth, organelle biosynthesis, and preparation for DNA replication	Cyclins (D-type), CDK4/6	Upregulation of cell cycle initiation genes (e.g., CYCD, CDK4)	It often occurs during the daytime and is linked to light cues	The G1/S checkpoint ensures sufficient size and absence of DNA damage	Critical for biomass accumulation and early developmental stages	Flow cytometry, transcriptomics, live-cell imaging
S Phase	DNA replication, synthesis of histones for chromatin assembly	Cyclins (E-type), CDK2	Activation of replication-specific genes (e.g., ORC1, MCM)	Follows G1; tightly regulated to ensure no overlap with mitosis	DNA integrity checkpoint monitors replication fidelity	Ensures genetic material is accurately duplicated for cell division	Pulse labelling with BrdU, FISH, and DNA content analysis
G2 Phase	Preparation for mitosis, synthesis of microtubules and other mitotic components	Cyclins (A-type), CDK1	Upregulation of mitotic preparation genes (e.g., CYCA, CDKB1)	Precedes the M phase, which photoperiods may influence	G2/M checkpoint ensures that DNA is fully replicated and undamaged	Vital for ensuring readiness for mitotic division and avoiding genomic instability	Western blotting for cyclin levels and microscopy of microtubule organisation

(continued)

Table 1 (continued)

Cell cycle phase	Key processes	Regulatory proteins	Gene expression	Time of occurrence	Checkpoint mechanisms	Significance in plants	Experimental methods
M Phase	Chromosome condensation, alignment, segregation, and cytokinesis	Cyclins (B-type), CDK1, Anaphase-Promoting Complex (APC)	Expression of mitotic spindle genes (e.g., KRP, MAD2)	It occurs rapidly and is independent of light–dark cycles	The spindle assembly checkpoint ensures proper chromosome alignment and attachment	Facilitates tissue development, organogenesis, and reproduction	Confocal microscopy, immunostaining for mitotic markers, and live imaging
G0 Phase	In the quiescent phase, cells exit the cycle for differentiation or dormancy	Low activity of cyclins and CDKs	Downregulation of cell cycle-related genes	It often occurs under stress conditions or in mature tissues	External signals like hormones regulate entry and exit	Allows cells to differentiate or conserve resources under stress	Gene expression profiling, hormone treatments, and cell viability assays

Sources [7, 13, 14]

CCA1 and LHY genes. These genes encode rhythmic transcription factors that peak in the morning and fall off by evening [7, 13, 14]. The proteins produced by CCA1 and LHY effectively inhibit the activity of TOC1. This gene has an evening peak and is considered to play a key role in the negative feedback loop, which supports and stabilizes the circadian cycle. Some reciprocally regulate CCA1, LHY, and TOC1, a characteristic of the plant circadian clock that ensures the rhythm is maintained no matter the time of day. The timing of expression of these genes and their interactions with other cellular processes is essential for regulating the internal responses to environmental signals in the plant, this includes the timing of cell division as well as the advancement of the cell cycle.

The relationship between the circadian clock and cell cycle is essential to plant growth and development. Cell Cycle in Plants The cell cycle phases in plants include G1, S, G2, and M, each involving specific events associated with cell growth and the replication of DNA, followed by cell division. They have evolved to align with the plant's internal circadian rhythm, ensuring cell division happens when it can most likely optimize its resources and respond efficiently to environmental conditions. Circadian rhythms establish cell cycle rhythms and regulate progression through the cell cycle, including DNA replication and the mitotic checkpoint. The cell cycle is an excellent example of this; it is known to be synchronized to a specific time of the day in plants, and DNA replication and cell division peak when plants are exposed to light in the daytime. This coupling of the cell cycles and the circadian clock, which enables the organism to grow while ensuring optimal cell division, is fundamental for life [15, 20, 21].

The mitotic cycle in root meristems is one of the most studied examples of the circadian clock and the cell cycle interaction. In plants, root meristems are cellular regions where cell division occurs (at the tips of plant roots), generating new cells that can effectively drive root growth. These meristems are extremely sensitive to internal and external signaling, including the plant's circadian rhythm. Studying the timing of mitotic events in root meristems, it was found to be closely related to the circadian clock. Specifically, it is preferentially during the daytime under optimal cellular conditions, such as light and energy for cell division, that the root meristem undergoes mitosis. Such synchronization enables a plant to optimize its growth and resource acquisition during the peak photosynthetic phase.

Circadian rhythms regulate the mitotic cycle in root meristems and have been shown to affect cell cycle regulation in various models, including *Arabidopsis thaliana*. In *Arabidopsis*, the oscillatory expression of CCA1, LHY, and TOC1 correlates with the timing of mitosis in the root meristem. They are thought to modulate the activity of genes that regulate the cell cycle, including cyclins and cyclin-dependent kinases (CDKs), which are major cytoskeletal regulators of cell division. Previous studies show that disruption of the molecular circadian clock can cause mistimed cell division in root meristems, leading to defective root growth and development. These results highlight the role of the circadian clock in regulating cell cycle timing and help ensure these events coincide with appropriate times for plant growth [2, 3, 23].

The regulatory machinery of circadian control of the cell cycle may differ between plant tissues; in other plant tissues, there are different circadian regulatory pathways.

For example, although the circadian clock is tightly integrated with the timing of cell division in the root meristem, it can also modulate cell division in tissues, including shoots and leaves. For example, in shoots, the circadian clock regulates cell cycle event timing to maximize growth and response to light. In these tissues, the circadian clock controls when cells will divide, synchronizing these processes to ensure that cells divide when the plant is in light and can use the energy produced by photosynthesis to grow leaves [2, 14, 23]. Here, the circadian clock controls when cells divide so that cells divide when the plant has access to light and other resources to exploit them most effectively. Circadian clock-cell cycle interactions in different tissues can help coordinate overall growth and development to prepare the plant to respond appropriately to varying conditions.

On a molecular basis, the circadian clock regulates the cell cycle by complex regulatory circuits with cyclins, CDKs, and other cell cycle modulators. Plant cell cycle control is a complex and highly regulated process involving the activity of different cyclins and cyclin-dependent kinases (CDKs) essential for cell cycle progression. These proteins, known as cyclins, bind to cyclin-dependent kinases (CDKs), activating their kinase activity and consequently phosphorylating target proteins that influence different events in the cell cycle, including DNA replication and mitosis—circadian control of cyclins, proteins which fluctuate over a day. The circadian clock controls the expression of particular cyclins, such as CYCD2 and CYCD3, to guarantee that the cell enters the accurate cell cycle stage at the right moment. Many other genes governing DNA repair, cell growth, and division are also under circadian control, alongside cyclins and CDKs. It regulates the cell cycle to coordinate with the plant's internal rhythms and environmental signals [10, 12, 14, 15].

There are also feedback loops mediated by proteins like TOC1, which play an essential role in synchronizing the rhythmic expression of cell cycle genes with the circadian clock. As a negative regulator, TOC1 represses the expression of CCA1 and LHY and regulates the expression of cyclins and CDKs. The rhythmic output of TOC1 is necessary for synchronizing the circadian clock with the cell cycle, and perturbations in TOC1 expression can result in mistimed cell division and growth defects; besides TOC1, other control proteins, including PRR5 (Pseudo-Response Regulator 5), PRR7, and GI (GIGANTEA), as well as various other regulators. The proteins help integrate signals from the environment, like light and temperature, with the plant's internal rhythms, ensuring cell division and growth happen at the right moment.

The circadian clock is an essential regulator of the timing of cell division in plants, so cells divide at optimal times in the 24-h cycle. The core circadian clock components, like CCA1, LHY, and TOC1, interact with cell cycle regulators, including cyclins and CDKs, to align cell cycle events with environmental cues. How circadian rhythms interact with the cell cycle is critical for coordinating between cell division and cell expansion to support efficient plant growth and development, and perturbation in this coordination leads to compromised growth and development defects. Various studies on model organisms, like *Arabidopsis*, have given us hints on how the circadian clock is tied to the cell cycle molecular pathways, and more will be investigated in the future. New insights regarding circadian rhythms and the cell cycle interplay can further enhance plant productivity, stress tolerance, and health.

4 Light as a Key Regulator of Cell Division

Light perception is central in regulating plant growth and development, influencing processes such as seed germination, phototropism, flowering, and cell division. Plants, being sessile organisms, rely heavily on light signals to synchronize their internal processes with the external environment. One of the most crucial functions regulated by light in plants is the cell cycle, particularly the process of cell division. This regulation is primarily mediated by photoreceptors, which allow plants to sense light and translate it into molecular signals that control cellular events [2, 3, 7, 23]. Among the key photoreceptors involved in regulating cell division are phytochromes and cryptochromes, which are sensitive to different wavelengths of light, including red, far-red, blue, and ultraviolet light. These photoreceptors control the timing of cell division by influencing various checkpoints in the cell cycle, most notably the G1-to-S transition, which is a critical point where the cell commits to DNA replication and division. The interplay between light perception and cell cycle progression helps optimize plant growth in response to environmental conditions, ensuring cell division occurs at the appropriate times and in coordination with other physiological processes, such as photosynthesis and nutrient acquisition.

Phytochromes are a family of photoreceptors that primarily perceive red and far-red light, allowing plants to assess the light environment and adjust their growth accordingly. Phytochromes function through a reversible photoconversion process, switching between two conformational states: the active form (Pfr) and the inactive form (Pr). The Pfr form, stabilized in red light, is primarily responsible for initiating light-induced gene expression that influences various cellular processes, including cell division. Phytochromes regulate cell division by modulating the expression of genes involved in the G1-to-S transition, a key point in the cell cycle. Upon exposure to light, phytochromes activate downstream signaling pathways that lead to the expression of genes encoding cyclins and cyclin-dependent kinases (CDKs), two major cell cycle regulators. The G1-to-S transition is controlled by the complex interplay of cyclins and CDKs, which phosphorylate target proteins to drive the cell into the S phase, where DNA replication occurs. Phytochromes, through their action on these signaling pathways, ensure cell division is synchronized with light availability, allowing the plant to maximize its growth potential during optimal light conditions [2, 8, 14, 24].

In addition to phytochromes, cryptochromes are another class of photoreceptors sensitive to blue and ultraviolet-A light. Cryptochromes regulate plant growth and development, including circadian rhythms, phototropism, and cell cycle progression. These photoreceptors contribute to the control of the G1-to-S transition by influencing the expression of cell cycle regulators. Like phytochromes, cryptochromes are involved in light-induced gene expression, but they typically have distinct effects, often interacting with other signaling pathways to fine-tune the timing of cell division. Cryptochromes modulate gene expression in cell cycle regulation by activating transcription factors and other downstream signaling molecules. The interaction between cryptochromes and the circadian clock also helps coordinate cell division timing with the day-night cycle, ensuring that cells divide when resources, such as light and energy, are most available.

The regulation of cell division by light perception is not limited to the action of photoreceptors alone; the timing of cell cycle events is also influenced by the diurnal patterns of light that plants experience throughout the day. Light is not a constant signal; it fluctuates in intensity and duration based on the time of day, season, and environmental conditions. These diurnal fluctuations in light intensity play a critical role in determining the rate of cell division, especially in meristematic tissues, where cell division is most active [7, 8, 10, 20]. The G1-to-S transition, for example, occurs more frequently during periods of higher light intensity, as plants can use the energy generated by photosynthesis to fuel the process of cell division. Conversely, during periods of low light intensity, such as at night, the cell cycle slows down or pauses, allowing the plant to conserve resources and prepare for the next cycle of division. This synchronization of cell division with the diurnal light cycle ensures that plants can maximize their growth potential when environmental conditions are favorable.

Meristems are regions of active cell division located at the tips of roots and shoots, and they are responsible for producing new cells that contribute to the growth and development of the plant. In many plants, meristematic activity is tightly coupled to the photoperiod, or the length of the day, which determines the timing of cell division. Studies have shown that in response to changes in photoperiod, meristematic tissues adjust their cell division rates to match the available light; for example, in long-day plants, which flower when the day length exceeds a certain threshold, cell division rates in the meristems increase during the day and decrease during the night. In short-day plants requiring shorter days to initiate flowering, cell division rates in the meristems follow a similar pattern, with an increase in cell division during the day and a decrease during the night. The ability of meristematic tissues to synchronize cell division with the photoperiod allows the plant to optimize its growth and reproduction, ensuring that new cells are produced at the right time and in response to the prevailing environmental conditions [15, 21, 24].

The molecular mechanisms integrating light signaling with cell cycle regulation are complex and involve multiple signaling pathways. Light-induced gene expression is central to this process, allowing the plant to activate or repress specific genes involved in cell cycle progression. Phytochromes and cryptochromes, through their interactions with transcription factors and other signaling molecules, regulate the expression of key cell cycle genes, including those encoding cyclins, CDKs, and other regulators. Cyclins and CDKs are crucial for driving the cell through the different phases of the cell cycle, and their activity is tightly controlled to ensure that the cell divides only when it is ready. The expression of these genes is influenced by light signals, ensuring that the timing of cell division is coordinated with light availability.

In addition to cyclins and CDKs, other proteins involved in cell cycle regulation, such as the retinoblastoma-related protein (RBR), also link light signaling to the cell cycle. The RBR protein regulates the progression of the G1 phase by interacting with E2F transcription factors, which are involved in activating genes required for the S phase. Through phytochromes and cryptochromes, light signals modulate the expression of genes encoding RBR and E2F, thereby controlling the timing of the G1-to-S transition. This regulation ensures cell division synchronizes with light availability, optimizing plant growth and development [2, 3].

Furthermore, integrating light signaling with the circadian clock adds a layer of regulation to cell division. The circadian clock controls the timing of various physiological processes, including the cell cycle, by regulating gene expression in cell division. Light signals, through their effects on phytochromes and cryptochromes, influence the circadian clock, ensuring cell division occurs at the right time of day. This coordination between light perception, the circadian clock, and the cell cycle is essential for ensuring that plant cells divide at optimal times in coordination with other processes, such as photosynthesis and nutrient uptake [2, 3, 8, 15].

Light perception plays a crucial role in regulating plant cell division by controlling the timing of key events in the cell cycle, notably the G1-to-S transition. Phytochromes and cryptochromes, the central photoreceptors in plants, mediate this regulation through their effects on gene expression and the activation of signaling pathways that influence cell cycle regulators such as cyclins, CDKs, and RBR. Diurnal light patterns also affect cell division rates in meristematic tissues, ensuring cells divide when environmental conditions are most favorable. The molecular mechanisms underlying light signaling and cell cycle regulation are complex and involve the interaction of multiple signaling pathways. Understanding how light perception integrates with the cell cycle is crucial for improving plant growth and productivity and developing strategies to optimize crop yield in response to changing environmental conditions (Table 2).

5 Temperature and Its Effect on Diurnal Cell Division Cycles

Temperature is one of the environmental variables that strongly affects the growth and development of plants (e.g., cell division). Cell division occurs in a highly regulated stepwise process through phases (G1, S, G2, and M), which must be timed, sequenced, and accurately signaled through molecular pathways. In the specific process of cell division, temperature controls the cellular activities of plants, such as metabolism, enzymatic activity, and gene expression related to the cell cycle and cell development; thus, it can be used as an excellent promoter of cell division in plants. Overall, the temperature effect on cell division isn't fixed; it modulates with the ISO time, and the dark vs light imparts differential responses to them [4, 20, 21, 24]. This diurnal modulation of cell division by temperature cycle is critical for plants to tune their internal environment with the external one. In addition, we might

Table 2 Light as a key regulator of cell division

Aspect of regulation	Light Type	Key photoreceptors	Regulated processes	Signaling pathways	Developmental context	Environmental adaptation	Experimental evidence
Phase-specific activation	Red, far-red	Phytochromes (PhyA, PhyB)	Activation of G1/S transition via cyclin D expression	Phytochrome-interacting factors (PIFs), HY5	Seedling development, early growth stages	Shade avoidance, photomorphogenesis	Mutant studies in Arabidopsis demonstrate PIF involvement; transcriptomics shows cyclin D expression
Circadian alignment	Blue	Cryptochromes (Cry1, Cry2)	Synchronization of cell cycle phases with light–dark cycles	CCA1/LHY circadian loop	Growth optimization during diurnal cycles	Efficient use of light resources in fluctuating environments	Time-lapse microscopy of dividing cells under controlled light conditions; RNA-seq of circadian-regulated genes
Microtubule organisation	UV, blue	Phototropins (Phot1, Phot2)	Regulation of microtubule dynamics during mitosis	MAPK signaling cascade	Root elongation and directional growth	Adjusting root architecture in low-light conditions	Confocal microscopy of microtubule arrays under UV/blue light; chemical inhibitors confirming pathway roles
Cyclin expression	Red, blue	Phytochromes, Cryptochromes	Light-dependent transcription of cyclin A and cyclin B	PIF-regulated transcription factors	Leaf and shoot growth phases	Rapid growth under favorable light conditions	Gene knockout experiments show loss of cyclin regulation under altered light regimes

(continued)

Table 2 (continued)

Aspect of regulation	Light Type	Key photoreceptors	Regulated processes	Signaling pathways	Developmental context	Environmental adaptation	Experimental evidence
Mitotic entry	Red, far-red	Phytochromes	Promotion of G2/M transition through CDK activation	COP1 suppression by active phytochromes	Floral meristem development	Photoperiodic flowering response	Protein assays showing CDK phosphorylation levels; light spectrum manipulation experiments
Chromosome alignment and spindle	Blue, red	Cryptochromes, Phytochromes	Regulation of spindle assembly and chromosome segregation during metaphase	Light-responsive mitotic checkpoints	Gametogenesis and fertilisation stages	High-fidelity chromosome segregation under changing light conditions	Live-cell imaging of mitotic spindles under varied light wavelengths; chromosomal analysis in mutant lines
Stress-induced cell cycle arrest	UV, blue	Phototropins, Cryptochromes	Arrest of the cell cycle under high UV light stress by inhibiting DNA replication	ROS signaling pathways	Defense mechanisms during abiotic stress	Prevention of DNA damage during high UV exposure	ROS quantification under UV stress; cell cycle marker analysis using flow cytometry

Sources [4, 20, 21, 24]

have to combine temperature rhythms with the organism's circadian clock to control cell division and optimize cell division timing in favor of growth and survival.

Plants encounter daily temperature fluctuations, typically higher temperatures during the day than at night. This diurnal regulation impacts many cell processes, but cell division is mainly affected by changes in temperature. Cell cycle regulation based on enabling transition between and progression through different cell cycle phases has been observed through diverse molecular networks and is temperature sensitive. For example, cyclins and cyclin-dependent kinases (CDKs), which drive the cell through the G1, S, G2, and M phases, are subject to temperature regulation, and environmental changes can influence their activity. Temperature can induce distinctly different growth and development arrest patterns, so the expression of cell cycle-related genes can be upregulated or downregulated depending on the temperature to assess the processes of cell division. Instead, cool temperatures allow but do not require slower cell cycles, lingering longer in each stage. Meanwhile, endpoint temperature can make the cell cycle run faster and the time to divide the cell faster. However, if the temperature is very high or low, the normal cell cycle will be broken, leading to reduced cell division and slowed plant growth. These effects underscore the necessity for adept temperature regulation to promote and preserve plant health.

In addition, part of the complexity of cell division in plants relies on the interplay between temperature rhythms and the circadian clock. The circadian clock is one of the internal timekeeping machineries that tunes an organism's physiological function to the 24-h day-night cycle. It controls the gene expression that drives the cell cycle to regulate many physiological processes, including cell division. The circadian clock is affected by environmental factors such as temperature, light, and nutrient levels, which allow the internal workings of the plant to adapt to its surrounding environment. In this way, temperature cycles influence the circadian clock, which controls the timing of cell division. The circadian clock can modulate the plant's response to temperature increases so that cell division occurs at optimal times of the day's light and dark parts [3, 7, 12, 23].

Under mild temperatures, the circadian clock synchronizes the timing of cell division with the plant's metabolic processes. During the day, when the temperature is higher and metabolic processes are at their peak, the plant has more energy and resources available to support growth, and cell division is generally more enforced. When it gets colder, maybe cell division is downregulated or stopped, and metabolism slows down in the dark to use energy sparingly for the plant. This regulation ensures that cells can only replicate themselves when sufficient resources are available, which promotes their proper growth and development. Extreme temperatures occur when temperatures deviate from the optimal range of most plants, and they may affect the circadian clock, which cannot synchronize cell division, and the disrupted cell cycle affects the growing plants adversely. For example, prolonged exposure to stress at temperature extremes can uncouple the phase of the plant's circadian clock from the cell cycle, disrupting the timing of cell division and potentially decreasing cell proliferation rates.

The temperatures accelerate and slow down the timing and efficiency of the cell division process. Cellular stress from high-temperature tissues led to common changes in pathway responses and cell cycle progression. Accumulation of reactive oxygen species (ROS) can cause heat stress and damage to cellular components, including proteins, lipids, and DNA. During such tissue injury, the plant uses heat shock proteins (HSPs) and other stress pathways to maintain cellular integrity. While these protective mechanisms allow the plant to endure heat stress, they also interfere with the normal progression of the cell cycle. The high temperature, however, can initially also stimulate the cell cycle rate, and accordingly, the rates of cell division increase. However, it is succeeded by a slowdown, or a stall, of the cell cycle as the plant mounts a stress response during the heat events. These phases can take a longer or shorter time, which causes uneven distribution of cells, leading to a lower yield of the plants [12, 20, 21].

However, low temperatures can also harm the process of cell division. Cold stress often affects normal cellular processes such as protein synthesis, enzyme activity, and cell cycle progression. When it is cold, the plant's metabolic rate drops, and cell division might stop or slow down to a crawl. The plant can also enter a dormant-like state, where the cells stop dividing to conserve energy. However, extended freezing can cause irreversible damage at the cellular level (e.g., disruption of membranes and denaturation of proteins), which would still inhibit progression through the cycle. Even mild cold stress can result in cells not dividing as quickly, so the plants may grow and develop more slowly.

In crop systems, the timing of cell division is a key prerequisite for productivity and yield, and each level of response to higher and lower temperatures will have a more drastic impact. Temperature-sensitive plants may grow stunted or yield less when subjected to ambient temperatures outside their critical range. For instance, high seasonally increasing temperatures can lead to accelerated cell division at an early stage of development. However, there may be slower rates of cell division because of heat stress, fewer cells, and a smaller plant size. Similar low temperatures during the growing season could also lead to a delay in cell division and a longer duration of the cell cycle, which would slow down growth and reduce crop yield. In both cases, synchronizing temperature rhythms with the circadian clock is crucial for the cell division cycle timing to coincide with optimal growth and productivity phases [16, 23, 24].

Scientists have been exploring warming conditions in model organisms, such as the ubiquitous model organism Arabidopsis thaliana, to gain insights into cellular divisions under different temperatures. These studies indicated that temperature modulates the expression of genes required for cell cycle control, including genes encoding cyclins, CDKs, and other relevant cellular regulators. In adapting to temperature changes, Arabidopsis cells tightly regulate their divisions through cellular circadian clocks invoked in phase with temperature-sensitive gene expression. Arabidopsis' initial response to high temperature is a slowdown that follows rapid cell cycle progression in cell division rate as the organism activates heat stress responses. Similarly, low temperatures stop cellular division, and the plant conserves energy and becomes protective. These findings show that temperature plays a role in the precision of cell cycle timing and highlight the necessity of adapting these crops to extreme environmental conditions for optimal growth and yield.

In plants, the timing and progression of the cell cycle are modulated by temperature, influencing cell division and its subsequent activities. However, cell division is temporally regulated by temperature-dependent, dynamic programs to allow the differentiation of cells when it is most beneficial to plant partitioning, synchronized with energy and resource allocation for increased fitness. Temperature cycles are more complex because they also affect the circadian clock; the circadian clock helps plant cells coordinate cell division with the day-night cycle. Excessive heat and cold can disrupt the divisions between cells, leading to bloating and reduced productivity, particularly in cultivated crops. Understanding how temperature modulates the timing of cell division and its integration with the circadian clock is fundamental to maximizing crop yield and plant health in fluctuating environmental conditions.

6 Hormonal Regulation of Cell Division in Diurnal Rhythms

Plant hormones play a key role in regulating cell division, helping ensure growth and development appropriately respond to environmental signals. The primary hormones regulating this process are toxins, cytokinin, gibberellins, and ethylene. Their effects can be modulated by the plant's circadian rhythms and environmental inputs such as light, temperature, and availability of nutrients. In cell cycle regulation, auxins and cytokinins are prominently involved, and their interaction shows control over different cell cycle stages. Gibberellins and ethylene, in contrast, have more specific roles in promoting or inhibiting growth processes that contribute to diurnally regulated cell divisions. By decoding these hormones' actions and interaction with the plant's internal clock, scientists understand how plants time their growth and development in an ever-changing environment.

So, now you probably understand plant hormones, which influence cellular processes, and a major one responsible for cell cycle regulation is auxin. They mainly encourage cell elongation and division and work chiefly in meristematic tissues like the root and shoot apical meristems. For example, auxins (most notably indole-3-acetic acid, IAA) are produced in regions of active growth (shoot tips) and are relayed to other plant regions to initiate developmental events. Hence, auxins trigger the expression of genes directly involved in the cell cycle progression, especially in the G1-to-S phase, during which a quiescent cell enters an active division phase. This is partly impacted by regulating cyclins and cyclin-dependent kinases (CDKs), the central cell cycle machinery. A critical aspect of auxin function is how it can help regulate the balance between cell division and cell expansion, allowing for consistent new tissue production in one part of the plant while permitting the expansion and elongation of cells in other parts [3, 4, 23, 24].

Another class of hormones, called cytokinins, has an antagonistic but complementary role relative to auxins in regulating cell division. Cytokinins are mainly responsible for promoting cell division in the meristems of shoots and roots. They promote the gene expression responsible for the G1/S transition of the cell cycle, allowing cells to divide in response to developmental signals. Cytokinins also regulate the cell division cycle, stimulating new cell production and facilitating new tissue growth. Unfortunately, the effects of cytokinins on cell division are not clear-cut; the balance of auxins frequently influences the ratio of cytokinins found in a given tissue. A high auxin: cytokinin ratio favors cell elongation, and a lower ratio favors division. I elaborate on the investment of auxins and cytokinins and how this delicate balance allows cells to divide only when needed safely, and in the precise tissue parts. It is fundamental to balance the signaling of auxins and cytokinins during plant development to regulate the timing of cell division in response to internal and external signals.

In addition, the regulation of cell division is further modulated through the interplay between plant hormones and circadian rhythms. The plant's circadian clock, which coordinates internal processes to the 24-h day-night cycle, modulates hormonal signaling pathways to optimize the timing of cell division. Thus, in the example of the circadian clock directly regulating the expression of key genes in auxin and cytokinin synthesis and degradation, the inducible accumulation of these hormones is likely not just to the right but at the right time for developmental regulation. In the light phase, with high photosynthetic activity and energy, auxins and cytokinins may be higher, leading to increased cell division in actively growing tissues. When energy availability is low, these hormone levels may drop at night, decreasing cell division. Thus, coordinated with the availability of resources, this diurnal regulation of hormone levels allows cell division to occur when it is most advantageous to the plant.

Gibberellins and ethylene also significantly regulate cell division in response to environmental signals, auxins, and cytokinins. Gibberellins are plant hormones that stimulate growth and cell elongation [7, 9, 10, 21]. Moreover, they are essential for maintaining normal tissue function, e.g., shoot apical meristem system, dormancy-break, and active growing distance. Gibberellins (GAs) also promote cell division by facilitating the transition from G1 to the S phase. Gibberellins also encourage cell elongation and tissue expansion, which do not directly impact cell division but are counterbalanced through cell proliferation and may indirectly influence the rate of this process. These compounds play a vital role in stem elongation, leaf expansion, flower opening, and fruit development, stimulating cell division and growth to foster plant development. Gibberellin levels show diurnal fluctuation, with observed peaks in the light phase when energy is more available for growth processes. Like other hormones, gibberellins are regulated by the circadian clock, which synchronizes their effects on cell division and growth with the plant's internal rhythms.

Ethylene is another plant hormone regarded as a growth inhibitor that regulates senescence and stress responses. The role of ethylene in cell division is complex, with its effect often depending on the particular developmental stage or environmental condition. Typically, ethylene acts as a negative regulator of cell division in specific tissues, such as during the ripening of fruits or senescence of leaves, while promoting division in particular stress responses. It was previously unveiled that the effects of Ethylene on the cell cycle are mediated via its impact on the expression of genes regulating the cell cycle, recycling the G1 to the S phase. Environmental signals (temperature, light, and availability of water) also modify the role of this hormone in plants' use of cell division, and ethylene levels increase in response to abiotic stressors [4, 5, 15, 21, 24]. Ethylene has a well-established role in cell division. Still, it is further complicated by the interaction with the circadian clock, where the internal pulse of time can regulate ethylene production as a function of changes in light and temperature over the day. It also enables the plant to align its internal processes with the outside environment, ensuring growth and development happen at the most favorable times.

The hormonal interactions among the auxins, cytokinins, gibberellins, and ethylene, induced by a circadian rhythm along with the environmental signals, allow for properly regulated cell division across the life cycle of a plant. Cell division is a key factor in plant growth, as it determines when new tissues are produced, when organs will develop, and the plant's size and shape. Hormonal regulation of cell division is critical, but disruptions can result in developmental defects such as stunted growth, abnormal organ development, and impaired stress responses. Thus, deciphering the interplay of plant hormones and between hormones and the circadian clock and environmental stimuli is vital for optimizing their growth and development, especially in agricultural contexts where environmental conditions may vary.

Plant hormones, including auxins, cytokinins, gibberellins, and gaseous ethylene, are central players in controlling cell division timing and responding to internal and external cues. The interplay between these hormones and the plant's circadian rhythms adds complexity to this regulation, enabling the alignment of cell division with the day-night cycle. Thus, cell division coincides with times of high metabolic activity and resource availability, coordinating growth and development [7, 10, 11, 21, 23]. Our understanding of the roles of various hormones in regulating plant growth and development has progressed from an initial focus on the role of individual hormones to the complex interplay of multiple hormones and their receptors within the context of circadian control of cell division under environmental conditions, recently hormonal signal has become the focal point of studies exploring the molecular mechanisms responsible for such regulation in the underlying circadian mechanism that controls cell division (Table 3).

7 Diurnal Regulation of Cell Division in Specific Plant Tissues

Meristematic tissues are specialized regions in plants responsible for active cell division and growth, mainly located in the root apical meristem (RAM) and shoot apical meristem (SAM). These tissues are essential for plant development, allowing continuous growth throughout the plant's life. Cell division in meristems is not constant but is tightly controlled by diurnal rhythms, meaning it follows a daily cycle influenced by environmental cues such as light and temperature. This time-based regulation

Table 3 Different hormones and phases of the cell cycle affected

Hormone	Cell type	Phase of cell cycle affected	Effect on cell division	Diurnal influence	Signaling pathways	Key genes/proteins	Experimental evidence
Auxins	Plant cells	G1, S	Promotes cell elongation and division	Enhanced activity during daylight hours	Auxin response factors (ARFs), TIR1 pathway	Cyclin D, CDKB	Observed in root meristems under light/dark cycles
Cytokinins	Plant cells	G2, M	Stimulates mitosis	Peak activity during midday	CRE1/AHK signaling, histidine kinase cascade	CYCD3, A-type cyclins	Increased division rates in shoot meristems
Gibberellins	Plant cells	G1	Enhances progression through G1	Activity influenced by circadian rhythms	DELLA proteins, GA-GID1 pathway	Cyclin A, MYB transcription factors	Rapid growth in stem tissues during the photoperiod
Abscisic Acid	Plant cells	G1, S	Inhibits division during stress	Higher levels during night periods	ABA receptors (PYR/PYL/RCAR), SnRK2 kinases	CDK inhibitors (ICK/KRP)	Reduced cell division under drought conditions
Ethylene	Plant cells	G2, M	Modulates stress-related division	Circadian regulation of ethylene production	EIN2, EIN3/EIL transcription factors	Ethylene-responsive genes	Altered division patterns in response to stress
Insulin	Animal cells	G1, S	Promotes growth and division	Regulated by feeding cycles	PI3K/AKT/mTOR signaling	Cyclin D1, CDK4	Enhanced proliferation in mammalian cell cultures
Glucocorticoids	Animal cells	G1	Inhibits cell division	Higher levels during early morning	Glucocorticoid receptor (GR) signaling	p21, p27	Observed in stress-induced models

(continued)

Table 3 (continued)

Hormone	Cell type	Phase of cell cycle affected	Effect on cell division	Diurnal influence	Signaling pathways	Key genes/proteins	Experimental evidence
Quorum-Sensing Signals	Microbial cells	All phases	Synchronizes division across populations	No diurnal influence in constant environments	Autoinducer signaling systems	LuxR, LuxI	Bacterial colonies with synchronized growth
Brassinosteroids	Plant cells	S, G2	Promotes cell elongation and division	Peak levels during daylight hours	BRI1/BAK1 signaling pathway	Cyclin B, CDK1	Enhanced division rates in leaf tissues
Melatonin	Animal cells	G2, M	Regulates the cell cycle at night	Higher levels during dark phases	MT1/MT2 receptor signaling	P53, cyclin B1	Reduced tumor growth in experimental models

Sources [4, 20, 21, 24]

ensures that cell division and growth are coordinated with the day-night cycle, optimizing resource use and developmental efficiency. As a result, plant growth aligns with external conditions, improving survival and fitness. A deeper understanding of how diurnal changes influence meristematic activity offers valuable insights into how plants synchronize their developmental processes with environmental rhythms, ultimately shaping their growth patterns and adaptive capacity [1, 5, 9, 18].

One key region responsible for root elongation is the root apical meristem (RAM), a zone of actively dividing cells located at the root tip. The RAM drives root growth by producing new cells, which enhances the plant's ability to absorb water and nutrients from the soil. Cell division in the RAM follows a diurnal rhythm, meaning it is synchronized with the daily light–dark cycle. While cell division tends to be more active during the day, coinciding with peak photosynthesis and metabolic activity, this pattern is not solely controlled by external light. Instead, it results from a complex interplay between environmental cues and the plant's internal circadian clock. Although light plays a significant role in regulating these rhythms through photoreceptor systems like phytochromes and cryptochromes, some plant species show root cell division even in the dark, albeit at a reduced rate. This indicates that root meristems can maintain rhythmic cell division independently of direct light exposure, guided by endogenous timing mechanisms. In this way, the circadian clock fine-tunes root growth, allowing plants to anticipate daily environmental changes and optimize resource uptake. Thus, root development is not simply a passive response to light but a regulated process that integrates external signals with internal biological timing to ensure efficient growth and adaptation across varying environmental conditions.

Diurnal rhythms also regulate shoot apical meristem (SAM) activity, although the impact of light and dark cycles on shoot cell division patterns may differ from root cells. The SAM generates the cells that will differentiate into the plant's leaves, stems, and flowers. It is also important to note that the shoot apical meristem (SAM), like the RAM, follows a diurnal pattern of cell division, with peak division rates occurring during daytime when ~~photosynthetic~~ energy is abundant. In all, by coupling cell divisions with daytime light, the plant can hypothetically introduce and transfer resources and bolster the growth during the period of highest energy input. The SAM is also influenced by the plant's circadian clocks, which contribute to the timing of development and synchronize with environmental shifts, such as light and temperature [10, 15, 21, 23, 24]. The plant response to ecological oscillation in light is mediated (directly or indirectly) by light-sensitive proteins like phytochromes and cryptochromes that allow the shoot meristem division activity to coordinate with broader environmental cycles.

Plants use light energy to produce chemical energy (ATP and sugars) through photosynthesis, which fuels growth and cell division. The timing of cell division in leaves is crucial for making new cells and expanding the leaf blade. Like meristematic tissues, leaf cells show a diurnal rhythm; cell division is highest during the daylight period, when photosynthesis is active and energy is abundant. This ensures that new cells are made when the plant has enough energy to support growth.

At night, when photosynthesis stops, the rate of cell division in leaves drops significantly. This coordination helps the plant use its energy efficiently, aligning cell growth with its metabolic needs. The process is regulated by plant hormones such as auxins and cytokinins, which respond to light and the plant's energy status to control how fast cells divide. Additionally, the circadian clock, the plant's internal daily rhythm, helps synchronize cell division with the day-night cycle, ensuring that growth happens at the most favorable times. This system allows plants to optimize growth by matching cell production with energy availability.

Cell division in the cambium, a meristematic tissue responsible for secondary growth, is tightly regulated by diurnal rhythms. The cambium, located between the xylem and phloem, produces new vascular tissues that enable the thickening of stems and roots, which is essential for efficient transport of water, nutrients, and sugars throughout the plant. Like other meristematic tissues, cambial cell division follows a daily pattern occurring more actively during the light phase and slowing down during the dark phase. This rhythmic activity ensures that cell division is synchronized with the plant's metabolic capacity, allowing cells time to recover and properly organize after division, particularly within the two distinct zones of the cambium: the fusiform initials and ray initials. This diurnal control arises from a combination of internal circadian signals and external environmental cues such as light and temperature [3–5, 21, 23]. These rhythms help coordinate cell cycle progression so that energy-intensive processes like mitosis occur when resources such as ATP and carbon skeletons from photosynthesis are most abundant. In addition, plant hormones play a key regulatory role: the balance between auxins and cytokinins, two major classes of plant growth regulators (PGRs), influences both the rate and timing of cambial cell division. Auxins, produced in the shoot apices and young leaves, promote vascular differentiation, while cytokinins, often synthesized in the roots, stimulate cell division. Their coordinated action ensures proper vascular development and functional integration of xylem and phloem. The synchronization of cell division across different meristematic regions, including the root apical meristem (RAM), shoot apical meristem (SAM), leaf tissues, and cambium, ensures coordinated growth throughout the plant. This integration allows for optimal use of metabolic resources, aligning growth with daily cycles of energy availability. For example, peak photosynthetic activity during the day provides the necessary energy and building blocks to support mitotic activity in both primary and secondary meristems. Thus, diurnal regulation of cell division in all these tissues is crucial for maintaining plant health, structural integrity, and adaptive growth under changing environmental conditions [11, 16, 25]. By aligning internal biological clocks with external cues, plants efficiently manage growth, resource allocation, and stress resilience.

Sensory perception and circadian rhythms play a vital role in regulating the timing of cell division in meristematic tissues, ensuring that plant growth is synchronized with daily environmental and metabolic cycles. This temporal control is observed in various growth zones, including the root apical meristem (RAM), shoot apical meristem (SAM), leaf tissues, and vascular cambium. In these regions, processes such as cell division, leaf expansion, and secondary thickening are coordinated with diurnal rhythms, aligning growth with periods of optimal energy availability such as peak photosynthesis during daylight.

For example, cell division in leaves and cambial tissues is more active during the day when photosynthetic activity provides the necessary energy and building blocks for growth. At night, these processes slow down, allowing the plant to conserve resources. This rhythmic regulation optimizes the use of metabolic resources like ATP and carbon compounds, enabling plants to grow efficiently and adapt to changing environmental conditions.

By fine-tuning growth with internal biological clocks and external cues like light and temperature, plants achieve sustainable development, balancing energy input with growth output to enhance survival and fitness [11, 16, 21]. Understanding the molecular mechanisms behind these diurnal rhythms not only reveals how plants adapt their growth to environmental cycles but also opens new avenues for improving crop productivity and resilience in agriculture.

8 Diurnal Regulation of Cell Division and Plant Development

Diurnal rhythms play a fundamental role in regulating cell division throughout a plant's life cycle, directly influencing key processes such as germination, leaf expansion, stem elongation, flowering, and organ development. Cell division is not random but is tightly synchronized with the daily light–dark cycle, ensuring that new cells are produced when energy from photosynthesis is most abundant. This temporal coordination is governed by the plant's internal circadian clock, which integrates external cues like light and temperature to align cellular activities with optimal times of day [3, 8, 16]. By doing so, plants maximize growth efficiency and adapt to changing environmental conditions.

During germination, timely cell division in the root and shoot apical meristems is essential for establishing seedlings. Diurnal regulation ensures that these divisions occur during favorable metabolic windows—typically in the light phase when ATP and carbon resources from photosynthesis are readily available. Disruption of this rhythm can delay seedling establishment and lead to weak or stunted growth, as cells fail to divide at optimal times [13, 15].

In the vegetative stage, leaf and stem growth are strongly influenced by diurnal patterns. Leaf expansion depends on cell division followed by cell enlargement, with the highest rates occurring during the day. This timing allows the plant to coordinate cell production with photosynthetic activity, ensuring sufficient energy and building blocks for growth. At night, the focus shifts to resource storage and cellular maintenance, supporting the next cycle of growth [25]. Similarly, stem elongation is regulated by diurnal rhythms, enabling plants to optimize their architecture for light capture and competitive advantage over neighboring vegetation [14, 20].

Cell division also underpins tissue differentiation and organ formation. During early leaf development, rapid cell division generates a pool of cells that later differentiate into specialized tissues such as xylem, phloem, and epidermis. Circadian signals tightly time this transition from proliferation to differentiation, ensuring that structural and functional maturation occurs in sync with developmental stages [8]. In roots, cell division peaks during the day to prepare for nutrient uptake during active photosynthesis, while at night, energy is redirected toward maintenance and storage.

Even reproductive development, including the formation of flowers, stamens, pistils, and petals, is coordinated by circadian regulation of cell division, ensuring proper timing for pollination and seed set [10, 17].

Overall, the diurnal control of cell division enables plants to optimize resource use, enhance stress resilience, and improve reproductive success. By aligning internal biological clocks with environmental cycles, plants achieve efficient, adaptive growth. Understanding the molecular mechanisms behind this regulation opens opportunities for breeding crops with improved growth timing, climate resilience, and higher yields.

9 Environmental Stress and Diurnal Control of Cell Division

Diurnal regulation of cell division plays a vital role in helping plants adapt to various biotic and abiotic stresses, such as drought, salinity, and extreme temperatures. Under stressful conditions, plants often slow down or temporarily halt cell division to conserve energy and maintain cellular integrity. This ability to enter a temporary cell cycle arrest is a key survival strategy that allows plants to redirect resources toward stress defense mechanisms such as producing protective proteins, accumulating compatible solutes, or enhancing antioxidant activity rather than investing in growth. The timing of these responses is coordinated by the circadian clock, which aligns internal biological processes with daily environmental cycles. This ensures that cell division occurs during the most favorable times, such as cooler nighttime hours during drought or high light periods for energy capture, maximizing growth efficiency while minimizing damage.

For example, under drought stress, plants adjust their cell division patterns to reduce water loss. Instead of dividing during the hot, dry daytime, many species shift division to the night, when humidity is higher and evaporative demand is lower. This re-timing allows for continued growth with minimal water use. Additionally, plants may reduce the overall frequency of cell division, conserving energy-rich molecules like ATP and carbohydrates, which are redirected to root growth or osmotic adjustment to improve water uptake [11, 14]. The circadian clock regulates key genes in the abscisic acid (ABA) signaling pathway—such as *ABI3*, *ABI4*, and *ABF2,* which control stress responses and influence cell cycle progression, ensuring that growth is synchronized with the plant's capacity to cope with stress [15, 21].

Similarly, salinity stress disrupts water uptake and causes toxic ion buildup, prompting plants to modulate cell division through circadian control. Genes involved in ion transport, like Salt Overly Sensitive 1 (SOS1), Na^+/H^+ exchanger (NHX1), and High-Affinity K^+ Transporter (HKT1), are rhythmically expressed, helping maintain ion balance and osmotic stability. Cell division in root and shoot meristems is often reduced under high salt conditions to prevent damage from ion toxicity, with activity shifted to periods of lower stress [10, 16].

Temperature extremes also impact cell division. During heat stress, excessive daytime cell division can lead to protein denaturation and DNA damage. Plants respond by delaying division until cooler evening or night hours, reducing metabolic strain. In contrast, cold stress slows metabolism and cell cycle progression. Still, the circadian clock helps optimize the timing of division to make the best use of limited energy resources, avoiding energetically costly processes when conditions are suboptimal [17, 23].

Studies in crops like maize and wheat show that drought alters the diurnal pattern of cell division, increasing nighttime activity to avoid peak stress [20]. In rice and barley, salinity and temperature stress similarly modulate division timing, demonstrating that circadian control enhances stress resilience across species. This plasticity allows plants to maintain growth under fluctuating conditions, a crucial adaptation in the face of climate change.

Understanding these rhythms opens new opportunities for climate-smart agriculture. By breeding or engineering crops with optimized circadian regulation, we can improve yield stability, reduce fertilizer and pesticide use, and enhance nutritional quality. For instance, aligning nutrient application with peak metabolic activity or harvesting at optimal times of day can improve shelf life and nutrient content. Integrating this knowledge with IoT sensors, machine learning, and precision farming enables real-time adjustments to growing conditions, supporting sustainable and resilient food systems [3, 25].

In summary, the coordination of diurnal rhythms and stress responses is a fundamental mechanism that allows plants to balance growth and survival. Harnessing this knowledge can lead to more resilient, productive, and sustainable crops in a changing world.

10 Conclusion

The circadian rhythm is a fundamental biological mechanism that connects plant development with environmental cycles, allowing plants to grow efficiently and adapt to changing conditions. This internal biological clock synchronizes cellular activities, especially cell division, with the daily light–dark cycle, ensuring that energy-intensive processes like DNA replication and mitosis occur when resources from photosynthesis are most available. Most cell division happens during specific time windows in the day, regulated by clock genes that control key cell cycle proteins such as cyclins and cyclin-dependent kinases (CDKs). In model plants like *Arabidopsis thaliana*, the circadian clock precisely regulates the transition from the G1 to S phase of the cell cycle, ensuring DNA replication occurs at the optimal time of day for maximum efficiency.

This coordination allows plants to conserve energy at night, when photosynthesis stops, and ramp up growth during daylight hours. Environmental cues such as light quality, temperature, and plant hormones fine-tune this timing, integrating external signals with the internal clock to optimize growth. As a result, plants can adjust their development to match daily and seasonal changes, enhancing their survival in unpredictable environments.

The rhythmic control of cell division plays a critical role in shaping plant architecture. It governs key processes such as root elongation, leaf expansion, flowering, and fruit development. For example, synchronized cell division in the shoot apical meristem determines how leaves are formed and arranged, influencing overall plant structure and yield. Similarly, in roots, the timing of cell division affects root length and branching, which are vital for water and nutrient uptake. Under stress conditions like drought or salinity, the circadian system helps plants slow down growth and redirect energy toward defense mechanisms, improving stress tolerance.

These insights have important implications for agriculture and food security, especially in the face of climate change. By understanding how the circadian clock controls cell division, scientists can develop crops that grow more efficiently, use resources better, and withstand harsh conditions. For instance, crops engineered to align their growth with favorable environmental windows could produce higher yields with less water or fertilizer.

Advances in genomics and CRISPR-Cas9 gene editing allow researchers to modify key clock genes such as *TOC1*, *LHY*, and *CCA1 to* fine-tune the timing of cell division and improve stress resilience. At the same time, high-throughput technologies like transcriptomics and live-cell imaging are helping uncover the molecular links between the clock and the cell cycle. Studying natural variation across plant species can also reveal how different ecologies have shaped circadian rhythms, offering valuable genetic resources for breeding.

Beyond genetics, agronomic practices like adjusting planting times, irrigation schedules, or light exposure can be aligned with a crop's internal clock to boost productivity. When combined with precision agriculture tools such as IoT sensors and machine learning, these strategies can lead to more brilliant, more sustainable farming systems.

In summary, the regulation of cell division by the circadian clock is essential for plant growth, development, and adaptation. Bridging molecular mechanisms with real-world environmental rhythms opens new pathways to improve crop performance, resilience, and sustainability, paving the way for innovative solutions in both plant science and agriculture.

References

1. Ahmad N, Naeem M, Ali H, Alabbosh KF, Hussain H, Khan I, Siddiqui SA, Khan AA, Iqbal B (2023). From challenges to solutions: the impact of melatonin on abiotic stress synergies in horticultural plants via redox regulation and epigenetic signaling. Scientia Horticulturae 321:112369. https://doi.org/10.1016/j.scienta.2023.112369
2. Ambali OA, Olajide PA, Adetuyi BO, Oni PG, Oyebanjo O, Ogunlana OO (2024) In: Shafi SM, Baloch FS, Chung YS, Martínez ARBT-M (eds) Chapter 3—Melatonin: a positive regulator of plant growth and development. Academic Press, pp 23–41. https://doi.org/10.1016/B978-0-443-13814-0.00003-X
3. Bharati AC, Prasad B, Mallick S, Masram DS, Kumar A, Saxena GK (2023) In: Verma C, Verma BDKBT-H (eds) Chapter 7—Animal and plant hormone. Elsevier, pp 151–175. https://doi.org/10.1016/B978-0-323-91684-4.00028-1
4. Cao S, Chen ZJ (2024) Transgenerational epigenetic inheritance during plant evolution and breeding. Trends Plant Sci 29(11):1203–1223. https://doi.org/10.1016/j.tplants.2024.04.007
5. Ebeed HT, Ahmed HS, Hassan NM (2024) Silicon transporters in plants: unravelling the molecular Nexus with sodium and potassium transporters under salinity stress. Plant Gene 38:100453. https://doi.org/10.1016/j.plgene.2024.100453
6. Greenwood M, Locke JCW (2020) The circadian clock coordinates plant development through specificity at the tissue and cellular level. Curr Opin Plant Biol 53:65–72. https://doi.org/10.1016/j.pbi.2019.09.004
7. Kocsy G, Tari I, Vanková R, Zechmann B, Gulyás Z, Poór P, Galiba G (2013) Redox control of plant growth and development. Plant Sci 211:77–91. https://doi.org/10.1016/j.plantsci.2013.07.004
8. Lim ASP (2021) Chapter 18—Diurnal and seasonal molecular rhythms in the human brain and their relation to Alzheimer disease. In: Swaab DF, Kreier F, Lucassen PJ, Salehi A, Buijs CN (eds) The human hypothalamus: anterior region. Elsevier, vol 179, pp 271–284. https://doi.org/10.1016/B978-0-12-819975-6.00017-0
9. Marothia D, Kaur N, Jhamat C, Sharma I, Pati PK (2023) Plant lectins: classical molecules with emerging roles in stress tolerance. Int J Biol Macromol 244:125272. https://doi.org/10.1016/j.ijbiomac.2023.125272
10. Martel J, Rouleau N, Murugan NJ, Chin W-C, Ojcius DM, Young JD (2024) Effects of light, electromagnetic fields and water on biological rhythms. Biomed J 100824. https://doi.org/10.1016/j.bj.2024.100824
11. Metcalf AJ, Boyle NR (2022). Rhythm of the night (and day): predictive metabolic modeling of diurnal growth in chlamydomonas. MSystems 7(4). https://doi.org/10.1128/msystems.00176-22
12. Myster J, Moe R (1995) Effect of diurnal temperature alternations on plant morphology in some greenhouse crops—a mini review. Scientia Horticulturae 62(4):205–215. https://doi.org/10.1016/0304-4238(95)00783-P
13. Pál M, Szalai G, Gondor OK, Janda T (2021) Unfinished story of polyamines: role of conjugation, transport and light-related regulation in the polyamine metabolism in plants. Plant Sci 308:110923. https://doi.org/10.1016/j.plantsci.2021.110923
14. Sharma A, Negi NP, Raina M, Supolia D, Mahajan A, Rajwanshi R, Gautam V, Bhagat B, Kakoria H, Kumar D (2022) Phytomelatonin: molecular messenger for stress perception and response in plants. Environ Exp Botany 201:104980. https://doi.org/10.1016/j.envexpbot.2022.104980
15. Srivastava D, Shamim M, Kumar M, Mishra A, Maurya R, Sharma D, Pandey P, Singh KN (2019) Role of circadian rhythm in plant system: an update from development to stress response. Environ Exp Botany 162:256–271. https://doi.org/10.1016/j.envexpbot.2019.02.025
16. Stevens B, Diels J, Vanuytrecht E, Brown A, Bayo S, Rujweka A, Richard E, Ndakidemi PA, Swennen R (2020) Canopy cover evolution, diurnal patterns and leaf area index relationships in a Mchare and Cavendish banana cultivar under different soil moisture regimes. Scientia Horticulturae 272:109328. https://doi.org/10.1016/j.scienta.2020.109328

17. Umen J, Liu D (2023) In: Goodenough TE (ed). Chapter 8—Cell cycle and circadian rhythms. Academic Press, pp 173–210. https://doi.org/10.1016/B978-0-12-822457-1.00002-9
18. Vo BT, Mas P, Johannes F (2024) Time's up: epigenetic clocks in plants. Curr Opin Plant Biol 81:102602. https://doi.org/10.1016/j.pbi.2024.102602
19. Wang C, Zhu S, Jiang X, Chen S, Xiao Y, Zhao Y, Yan Y, Wen Y (2023) Spatio-temporal variation of species richness and phylogenetic diversity patterns for spring ephemeral plants in northern China. Glob Ecol Conserv 48:e02752. https://doi.org/10.1016/j.gecco.2023.e02752
20. Wang J, Huang J, Bao J, Li X, Zhu L, Jin J (2023) Rice domestication-associated transcription factor PROSTRATE GROWTH 1 controls plant and panicle architecture by regulating the expression of LAZY 1 and OsGIGANTEA, respectively. Molecul Plant 16(9):1413–1426. https://doi.org/10.1016/j.molp.2023.08.011
21. Wang Q, Li Y, Lin D, Feng X, Wang Y, Wang T, Ding H, Zhang J (2024) A growth-regulating factor 7 (GRF7)-mediated gene regulatory network promotes leaf growth and expansion in sugarcane. Crop J 12(2):422–431. https://doi.org/10.1016/j.cj.2024.01.006
22. Yan R, Ho C-T, Zhang X (2022) Modulatory effects in circadian-related diseases via the reciprocity of tea polyphenols and intestinal microbiota. Food Sci Hum Wellness 11(3):494–501. https://doi.org/10.1016/j.fshw.2021.12.007
23. Yun H, Lim S, Kim YX, Lee Y, Lee S, Lee D, Park K, Sung J (2018) Diurnal changes in CN metabolism and response of rice seedlings to UV-B radiation. J Plant Physiol 228:66–74. https://doi.org/10.1016/j.jplph.2018.02.007
24. Zhang J, Xiao C, Duan X, Gao X, Zeng H, Dong R, Feng G, Ma K (2024) Species' geographical range, environmental range and traits lead to specimen collection preference of dominant plant species of grasslands in Northern China. Plant Div 46(3):353–361. https://doi.org/10.1016/j.pld.2024.02.001
25. Zheng X, Zhang K, Zhao Y, Fent K (2021) Environmental chemicals affect circadian rhythms: an underexplored effect influencing health and fitness in animals and humans. Environ Int 149:106159. https://doi.org/10.1016/j.envint.2020.106159

Chapter 9
The Pulse of Pollination: Bees and Their Role in Plant Reproductive Rhythms

Abhishek Rajpurohit, Seema Yadav, and Manish Sarwa

1 Introduction

1.1 Overview of Plant Reproduction and Pollination

Plants use multiple forms of reproduction including pollination which serves as a basis for the reproductive cycle and seed production of flowering species [56]. Flowering plants predominantly use insects to achieve successful reproduction despite wind/air and water among other abiotic agents (91.5% according to [58]). Bees of Apidae-family Hymenoptera represent the most impactful and significant pollinators because their flower-cantered behaviours, strong loyalty to flowers and their ability to transfer pollen over extensive distances make them effective [43].

Bee colony activity leads directly to reproductive achievements in plants and simultaneously drives agricultural crop production upward. The pollination of 75% of world food crops depends on insects yet bees perform the primary role in creating fruits as well as producing seeds [28]. Scientific research and conservation work focuses on bees because their activities with diverse plant communities ensure ecosystem stability and plant biodiversity.

A. Rajpurohit (✉)
Department of Life Sciences, Lachoo Memorial College of Science and Technology, Jodhpur, Rajasthan, India
e-mail: abhiraj_rajpurohit@yahoo.com; abhishekrajpurohit@lachoomemorial.org

S. Yadav · M. Sarwa
Department of Zoology, Jai Narain Vyas University, Jodhpur, Rajasthan, India

A. Kothari Chhajer and V. Singh Gour (eds.), *Introduction to Rhythms in Plants*, Studies in Rhythm Engineering, https://doi.org/10.1007/978-981-95-3592-7_9

1.2 The Concept of Rhythms in Plant Reproduction

Plant reproductive operations function based on biological system cycles which control both flowering schedules and nectar availability and pollen dispersal period. Plant rhythms consist of circadian rhythms for daily cycles together with seasonal rhythms caused by photoperiod and temperature changes and developmental rhythms that manage pollination and fertilization timing [69]. The timing of flowers to open combined with nectar production happens through circadian rhythms to provide rewards as pollinators become most active [72].

Through evolutionary adaptation, plants alongside their pollinators developed matched life cycles to reach their finest pollination outcomes. Flowering plants such as certain flower species open and release sugary nectar during particular daily intervals which align with peak foraging hours of their chief pollinating bees [22]. The optimized communication improves the chances of pollen transportation that results in successful fertilization which produces seeds.

1.3 Purpose of the Chapter

The central focus of this chapter evaluates the role of bees as they contribute to rhythmic plant reproductive processes for pollination operations and survival improvement. The article investigates the relationship between bees and plant biological patterns while it reveals key insights into these relationships. What patterns bees follow during their foraging activities match plant reproductive phases. The role of circadian and seasonal rhythms in pollination success. The ecological and agricultural importance arises from maintaining proper alignment between plants and their pollinators.

These connections between biological rhythms of plants and bees require careful analysis because they enable development of protection approaches against climate change impacts and habitat destruction and honeybee population decline. Future sustainability of global food security along with biodiversity depends on thorough studies of bee-plant synchronization because pollination rhythms are now under threat due to environmental stressors [61].

2 Plant Biological Rhythms and Reproductive Timings

The operation of biological rhythms in plants actively controls their developmental cycle and their success in reproduction along with growth patterns. A range of self-governed biological timers inside plants enables them to match their physical processes and breeding cycles with outside signs such as light changes, temperature fluctuations and natural pollinators.

2.1 Circadian Rhythm in Plants

The internal body clocks called circadian clocks control the 24-h biological cycles which drive circadian rhythms. Regulatory rhythms in plants control four essential physiological procedures which include flower opening and scent release together with nectar liberation and leaf adjustments. Throughout time, many flowering plants have developed a strategy to open their flowers during particular day hours for attracting pollinators by releasing scent. *Nicotiana attenuata* (wild tobacco) releases its floral scent majorly during night to attract moth pollinators according to Yon et al. [78]. The evening bloom of *Ipomoea alba* (moonflower) attracts hawkmoths which are active during night time [65]. Numerous plant species generate nectar through time-dependent cycles which match the activity pattern of their essential pollinators. The plant species *Brassica rapa* regulates nectar secretion through its response to daylight patterns in order to optimize pollinator visits when they are most active [44]. The exact timing between plant flowers and pollinators leads to successful pollination processes which increase reproductive outcomes.

2.2 Seasonal and Environmental Rhythms

In addition to regular daily patterns, plants demonstrate season-based and environmental-based rhythms which control the timing of reproduction events. The flowering time of plants together with pollinator interactions depends heavily on photoperiod length and temperature patterns and climatic changes in the environment.

Plants possess photoperiodism as their ability to detect changes in the day duration which triggers appropriate floral development. The flowering behaviour of *Arabidopsis thaliana* as a long-day plant occurs during spring and summer longer days until autumn whereas *Chrysanthemum morifolium* blooms under short-day conditions during fall [2].

The pollination success together with floral timing depends on temperature variations. Plants experience faster blooming under warmer conditions but heat that is too intense will affect their flowering process and drive away essential pollinators. The reproductive success of plants in alpine ecosystems has suffered due to temperature increase which created pollination timing mismatch with their native pollinators [47].

2.3 Coordinated Timing for Successful Pollination

Plants alongside their pollinators evolved parallel adaptations which secure that flowering periods correspond to optimal pollinator abundancy. Plants from many species activate their blooms in harmony with pollinators when they become active.

The flower *Lotus corniculatus* reaches blooming stage during peak bumblebee active time based on findings by Ollerton and Lack [57]. Plants employ several different approaches to guarantee successful pollination through *Helianthus annuus* (sunflower) along with multiple other plant species use mass flowering by producing abundant flowers during compact time spans to draw pollinators towards their blooms.

According to Forrest and Miller-Rushing [26] studies on early-flowering species such as *Mertensia fusiformis*, it was revealed that persistent pollinator deficits, even during years with early snowmelt, highlighting a mismatch between plant phenology and pollinator availability. Interestingly, the pollinator assemblage visiting *M. fusiformis* changes predictably over the course of the flowering season, which may have important implications for natural selection on floral traits, especially in years when pollen limitation is significant. These findings suggest that early- and late-flowering populations may undergo differential evolutionary pressures driven by shifts in pollinator community phenology, potentially leading to divergence in floral morphology and reproductive strategies.

3 Bee Pollinator Behaviour and Timing

Bees ensure healthy reproduction in plants through pollination since plants display matching biological rhythms. Various contextual elements such as ecological signs from the environment and floral resource abundance along with innate species-stereotyped daily patterns affect both their movement patterns and their timing.

3.1 Circadian Rhythms of Bees

Bee foraging behaviour follows circadian rhythm patterns which establish biological timing between bee activities and peak flowering hours [54]. Each bee species shows its own schedule of activity which depends on environmental factors and their physical responses to these elements.

Honeybee (*Apis mellifera*) individuals display a robust circadian timekeeping system during foraging behaviours because they begin foraging activities early in the morning before their activity reaches its peak midpoint in the morning period. Dyer [21], honeybee foraging activity typically peaks in the mid-morning, with both temperature and light intensity playing key roles in regulating their foraging patterns, which tend to decline during the late afternoon.

Bumblebees (*Bombus* sp.): Bumblebees forage throughout most daytime hours extending to early mornings and evenings because their thermoregulatory capabilities exceed those of honeybees as reported by Heinrich [36].

The *Andrena* and *Megachile* species together with other solitary bees display unique periods for foraging activities where separate groups show preference either for the morning hours or those at midday [70]. The various foraging times among pollinators provide stable pollination services to different ecological environments.

Scientists have established that bees adapt their biological time mechanisms to flower food schedules as an adaptive way to reach maximum efficiency in foraging activities [7].

3.2 Flower Preferences and Timing

The flower selection preferences of bees develop from their natural biological time patterns combined with acquired knowledge about foraging resources. The reward consistency of predictable nectar and pollen found in flowers gets strengthened with habitual visits [16].

Plant species, in turn, establish flower-opening schedules (anthesis timing) that align with the daily activity patterns of their primary pollinators to optimize pollination success [73]. The flowers of morning-glory (*Ipomoea purpurea*) open at dawn, synchronizing with the early foraging of honeybees, while evening primrose (*Oenothera biennis*) opens its flowers at dusk, attracting nocturnal hawkmoths such as *Manduca sexta* [65].

Through associative learning, bees retain memorized information regarding when particular flowers will provide rewards. Scientific research involving honeybees shows these insects develop the ability to predict nectar accessibility which helps them return to flowers during their native time periods even when placed in controlled environments devoid of external time indicators [53].

Bees also adjust their foraging routines in response to environmental factors such as temperature, humidity and competition with other floral visitors. In arid and semi-arid habitats, solitary bees (e.g., *Centris pallida*) show peak activity during the cool early morning hours to avoid overheating and desiccation, while modifying their foraging range to track floral resources more efficiently [70, 75].

3.3 Diversity of Bees Associated with Plant Pollination

Bees are one of the most diverse and vital pollinators found on Earth, with over 20,000 known species worldwide, classified into seven major families: Apidae et al. [52]. This group includes both generalist pollinators, like the familiar honey bees (*Apis*) and bumble bees (*Bombus*), which visit a wide variety of flowering plants, and specialist bees, many of which are solitary and tend to stick to specific kinds of flowers. These specialist bees, called oligolectic, often forage from a single plant genus or family [18], which boosts their efficiency and forges strong bonds with those plants. For instance, *Macropis* bees are tightly linked with oil-producing *Lysimachia* plants, while *Eucera* bees frequently visit legumes such as *Lathyrus* and *Vicia*.

However, today's fast-changing environment is affecting this delicate balance. With climate change and habitat loss altering when and where plants bloom, bees are adjusting their behaviour too. Many species are shifting their foraging zones to higher elevations or more temperate regions, or they are starting to use new types of flowers that weren't part of their original diet [13]. These changes can disrupt long-established plant-bee interactions and affect pollination success. That's why protecting the diversity of bee species—from generalists to specialists—is crucial for maintaining healthy ecosystems and supporting agriculture around the globe.

3.4 Evolution of Bees with Flowering Plants and Shifts in Nectar Foraging

The story of bees is closely tied to the rise of flowering plants. Scientists believe that bees first appeared around 100–120 million years ago, during the Early Cretaceous period, evolving from wasp-like ancestors that originally hunted other insects [14]. As flowering plants—also known as angiosperms—began to spread and diversify rapidly, bees transitioned to a diet based on nectar and pollen. This shift led to the development of remarkable adaptations in many bee species. For example, *Euglossa* bees evolved long tongues to reach nectar in deep flowers, while *Bombus* (bumble bees) developed the ability to perform buzz pollination, shaking flowers to release pollen [76]. Over time, such close interactions gave rise to deep mutual relationships between specific plants and their bee pollinators, where each depended on the other for survival and reproduction.

But in today's world, things are changing. Human-driven pressures like climate change, habitat destruction and the spread of invasive species are altering the availability and timing of floral resources. As a result, many bees are being forced to adapt by changing where and how they forage. Some are moving to cooler, higher elevations or shifting towards the poles to follow the plants they rely on [40]. Others are adopting more generalist diets, feeding on whatever flowers are available—even if those aren't their usual partners—especially in cities or fragmented landscapes [5]. These shifts can break down the finely balanced relationships between bees and plants, potentially lowering pollination success, affecting seed production and weakening bee nutrition. While the long evolutionary journey of bees and flowers showcases a brilliant example of co-evolution, the disruptions we are witnessing today threaten this ancient alliance and raise serious concerns for the future of biodiversity and global food systems.

3.5 *The Role of Bees in Pollination Networks*

Different ecosystems depend on bees to act as their keystone pollinators which support the periodic movement of pollen between plant species. The patterns bees follow in their movements determine both plant reproductive outcomes and the dimensions of their genetic base.

Bees perform cross-pollination tasks by moving pollen from the same species of flowers and different species for the purpose of enhancing genetic diversity [50]. Plants in fragmented ecosystems need pollinators for maintaining reproductive linkage among their population.

The amount of pollination support received by flowers varies among different bee species. The pollinating services of honeybees include broad plant flower visits but bumblebees mostly carry out buzz pollination to cross-pollinate plants including tomatoes (*Solanum*) [11].

Bee populations featuring diverse schedules when they pollinate each other create uninterrupted pollination cycles spanning daily periods and seasonal timeframes [45]. Ecological stability and maintenance of plant-pollinator relationships occurs through redundant populations.

4 Synchronization Between Bees and Plants

The biological process of pollination works only through the exact synchronization of bees with plants. Through this partnership, both plants achieve successful reproduction while bees receive their necessary food. Bee foraging exists in precise timing because of evolutionary forces that have created beneficial relationships between bees and plants alongside adaptive patterns and complex life cycles [10, 23].

Flowering plants move through different linked stages from developing flowers to creating nectar and pollen to drawing pollinators to finally transferring pollen to achieve fertilization and form seeds and fruits. The stages of pollination intimately connect to environmental cues like temperature together with daylight duration and seasonal changes to maintain plant-pollinator synchronization [64]. The complex connection between plants and their pollinators exists through the essential steps shown in Fig. 1 which demonstrates plant reproduction along with forager actions. Visual representation in the diagram shows external elements that affect pollination periods alongside affecting its efficiency because of temperature variations and environmental alterations and farming practice modifications. Ecosystem stability and agricultural productivity face extensive consequences when disturbances affect any step within this mutual dependency between plants and pollinators.

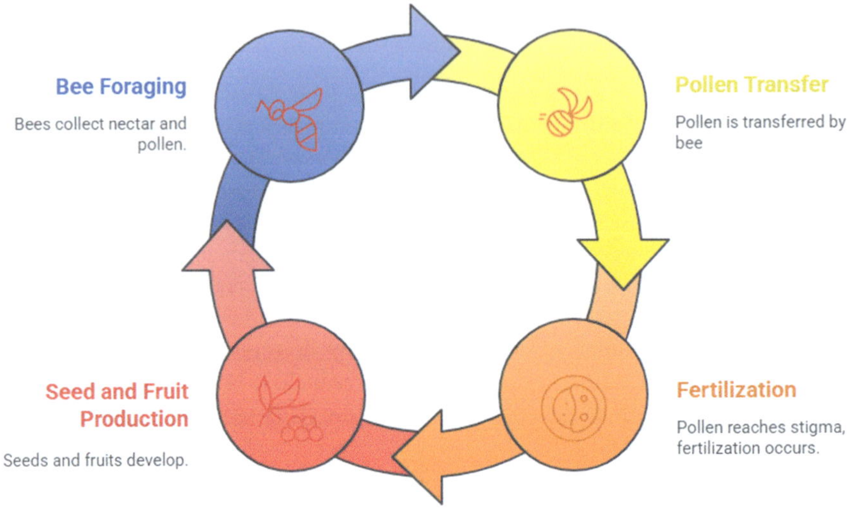

Fig. 1 Pollination cycle

4.1 Mutualistic Relationships

Bees maintain a helpful interaction with flowering plants by exchanging food supply from plants with their service to pollinate them. The connection results in biodiversity development and enhances ecosystem stability [58].

Plants use three main features including colour together with scent and nectar creation schedules as specific characteristics which attract their main bee pollinators. Honeybees (*Apis mellifera*) demonstrate increased attraction to blue and yellow flowers because the insects display strong visual attraction towards these colours according to Chittka and Menzel [15].

Plants that adapt to nectar production times match the active periods of their essential pollinator species. Morning-blooming flowers produce nectar at peak foraging times because it coincides with when honeybees and bumblebees are most active according to Fenster et al. [23].

Selected plant species develop complex relationships with exact bee populations to achieve pollination. The orchid bee (*Euglossini*) maintains a pollination relationship with specific orchid species that produce scent chemicals detected exclusively by those bees [10].

Plants benefit from optimal pollination support through this mutually beneficial relationship which bees use to obtain necessary colony nutrients.

4.2 Co-evolution of Timing

Time and again throughout evolutionary history bees and plants have evolved synchronized biological periods which lead to optimized reproductive gains. The evolutionary interaction leads to accommodative changes between plant flowering times and bee foraging patterns [4].

Plants developed specific flowering periods that match the maximum abundance of their essential pollinators during a particular time. Early blooming occurs in temperate spring ephemerals because it enables them to exploit newly emerged solitary bees according to Thomson [71].

Bees demonstrate plastic behaviour through their foraging schedules since they modify their activity according to seasonal together with daily transitions in flower abundance. Honeybees use their learning abilities to record the rewarding flower visitation periods thus maximizing their resource collection [15].

But this synchronization is getting affected by climate change. Plant phenology is changing due to temperature and photoperiod changes; many species are now blossoming earlier than they did in previous decades. Rafferty et al. [64] shown that warmer temperatures have accelerated the beginning of flowering in plants including bluebell (*Mertensia virginica*) and wild geranium (*Geranium maculatum*) in certain environments, such as alpine or temperate woodland settings. However, phenological mismatches may result from pollinator emergence shifting at a different rate, which is frequently triggered by distinct environmental factors like photoperiod or soil temperature. These discrepancies lessen the overlap between pollinator activity and flowering, which in turn lowers pollination success and may lead to decrease in bee nutrition and plant reproduction. A stable pollination system requires all adaptive strategies to maintain their alignment over time.

4.3 The Role of Bees in Flower Pollination Rhythms

The activity of bees serves two functions because they aid pollination and modify plant reproductive success by controlling both when fertilization happens and seed generation occurs [3]. Plant reproductive results are determined through three factors which include bee visit frequency and timing and visit efficiency.

After bees transfer pollens, they start fertilization processes which immediately influence the development of seeds and fruits. Frequent bee interactions play a vital role in self-incompatible plants because they guarantee the adoption of cross-pollination according to Fenster et al. [23].

Higher seed set results from increased bee activity which also promotes genetic diversity of seeds. The pollination services of multiple bee species result in better seed quantity and higher seed viability within almonds (*Prunus dulcis*) and sunflowers (*Helianthus annuus*) crops [4].

Floral lifetime gets affected by the number of bees visiting the blossoms. The receipt of proper pollination signals in several species results in flowers closing earlier to provide more resources towards seed development [71].

Plant reproductive achievements function in a complex manner with bee pollination interactions that control their success. Evolutionary processes led to the bee-plant relationship because plants and bees developed deep interconnected strategies that synchronize their behaviours to achieve successful pollination. Pollinator conservation with maintenance of ecological systems demands understanding these natural connections according to Burkle et al. [13].

5 The Ecological Significance of Synchronization

Bees and plants persisted through evolution because mutualistic connections and adaptive responses function together with timed pollination cycles for reproductive accomplishment. Pollinator conservation with maintenance of ecological systems demands understanding these natural connections according to Burkle et al. [13].

5.1 Ecological Impact of Timely Pollination

The timing of pollination acts as a critical component for maintaining ecosystem health and supporting biodiversity. The synchronization between pollinators and plants enables successful reproduction and genetic diversity which maintains ecosystem resilience [3].

Biodiversity and Genetic Variation: The synchronization of pollination processes facilitates cross-pollination between plants which boosts genetic diversity. Genetic variation among plants enhances their ability to adapt to environmental changes and various stressors including climate fluctuations and habitat destruction [23].

Ecosystem Stability: Timely pollination maintains food chain relationships because numerous species rely on flowering plants as their food source. Changes in pollination schedules result in widespread ecological disturbances that modify food web dynamics while reducing plant fertility according to Thomson [71].

Resilience to Climate Change: Pollination synchronization is essential for ecosystem resilience. Changes in flowering periods or pollinator behaviours due to climate change create timing mismatches which impair plant reproduction and decrease pollination success [64]. Species interactions and ecological stability depend on the preservation of this synchronization.

5.2 Benefits for Crop Production

The success of agricultural ecosystems relies greatly on synchronized pollination which is crucial for crops needing insect pollinators like bees [4]. Proper synchronization maximizes crop yields and quality.

Pollination by bees plays a vital role in boosting yields of insect-pollinated crops. For example, almonds (*Prunus dulcis)* are entirely dependent on insect pollination—particularly honey bees (*Apis mellifera*)—for fruit set. A study by Klein et al. [43] found that in almonds, coordinated bloom-time bee visits can increase nut set by over 20%. Similarly, apples (*Malus domestica*) benefit from synchronized pollination by both honey bees and wild bees such as *Andrena* spp., resulting in increased fruit set and better seed development. Blueberries (*Vaccinium corymbosum*), which require buzz pollination, see significantly improved yields when pollinated by bumble bees (*Bombus impatiens*), as these bees are more efficient at extracting pollen than honey bees [28, 39].

Pollination not only affects the quantity of produce but also influences quality parameters like size, shape, shelf-life and nutritional value. For instance, in strawberries (*Fragaria* × *ananassa*), bee-pollinated fruits are larger, more symmetrical and have a longer shelf-life compared to wind or self-pollinated ones. Klatt et al. [42] demonstrated that bee pollination improved commercial quality in strawberries by enhancing uniform development of the fruit due to optimal fertilization of ovules. Similarly, in oilseed rape (*Brassica napus*), synchronized bee pollination leads to higher seed oil content and better seed set [8].

The requirement for some agricultural inputs, particularly artificial fertilizers and human pollination, can be decreased with effective and consistent pollination. Although fertilizers mainly provide nutrients to the soil, they are unable to replace effective pollination, which is necessary for the development of fruit and seeds. Better pollination, however, increases the efficiency of nutrient usage. Improved pollination, for instance, enables more effective distribution of plant resources to growing seeds and fruits in cucumbers (*Cucumis sativus*) and sunflowers (*Helianthus annuus*), lowering the excessive use of nitrogen fertilizers [1].

Furthermore, by reducing the ecological impact of farming, dependence on consistent pollination services—especially from wild bee communities—promotes sustainable agriculture. According to Potts et al. [61], implementing pollinator-friendly techniques (such as lowering pesticide use and planting hedgerows) improves agricultural output and biodiversity, which in turn lessens the need for chemical or mechanical inputs over the long run.

5.3 Sustainability and Pollination Services

Healthy pollination networks are essential for both food security and long-term ecological sustainability.

Pollination as an Ecosystem Service: Bees play a major role in providing pollination services worldwide, which promote agricultural productivity as well as the reproduction of wild plants. According to studies, animal pollination helps more than 75% of the world's food crops [58].

Sustainable Agriculture Methods: Pollination services will continue if pollinator habitat conservation, less pesticide usage and pollinator-friendly agricultural landscapes are implemented. Agroforestry and wildflower corridors are two pollination-friendly techniques that support pollinator-crop synchronization [28].

Global Stability and Food Security: Pollination scheduling errors can result in lower agricultural production, which impacts food supply chains and financial stability. Maintaining ecological balance and long-term food production requires sustainable pollination management [43].

6 Disruptions in Pollination Rhythms

For healthy reproduction and the stability of an ecosystem, pollinators and plants must synchronize. However, this synchronization is disrupted by a number of anthropogenic and environmental causes, including pollinator reduction, temperature change and timing mismatches, which results in ecological imbalances and decreased pollination success [12, 64].

6.1 Climate Change and Its Effects

The environmental cues that govern interactions between plants and pollinators are drastically changed by climate change. Temperature increases, changes in precipitation patterns and severe weather conditions affect when plants bloom and when bees are active [51].

Temporal mismatches result from pollinators' inability to change their activity at the same rate as plants blooming earlier in the season due to rising temperatures [48]. For instance, a study on the Japanese flowering plant *Fritillaria camschatcensis* revealed that early blossoming brought on by warming reduced pollination rates because bumblebees were less active [48].

By changing their energy consumption and impairing their capacity to find flowers at the appropriate moment, extreme heat can decrease bees' foraging activity [68]. In desert areas, where high temperatures restrict their foraging windows and lower pollination efficiency, this is especially noticeable for stingless bees (*Trigona* sp.) [63].

While some plants depend on the length of the day (photoperiod sensitivity), others rely on temperature cues to start blooming. Decoupling of these variables due to climate change may increase the likelihood of pollination failure [35].

6.2 Pollinator Decline

Pathogens, pesticide exposure, habitat loss and climate change are all contributing factors to the decline in bee populations [61]. Synchronous interactions between plants and their pollinators are adversely affected by the decline in pollinator numbers.

Pollen transport and transfer efficiency decreases when pollinator abundance declines because fewer flowers are visited. For instance, Garibaldi et al. [28] discovered that, even in the presence of honey bees (*Apis mellifera*), pollination-dependent crops like coffee (*Coffea arabica*), mango (*Mangifera indica*) and sunflower (*Helianthus annuus*) showed a 25–40% decrease in fruit set in fields with low wild pollinator visitation in a number of global regions, including parts of Latin America, Europe and Asia. These results demonstrate that the losses of wild pollinators cannot be entirely offset by managed bees alone.

Certain bee species are necessary for the efficient pollination of many plants. Reduced seed production, decreased genetic variety and ultimately population reduction result from the reproductive success of dependent plants being jeopardized by the decline of these specialist bees. For instance, the solitary bee (*Andrena nigroaenea*) is necessary for the pseudocopulatory pollination of the orchid (*Ophrys sphegodes*). This orchid's inability to reproduce is directly caused by a drop in the bee population [6].

Decreases in wild bee species, especially solitary bees like *Osmia lignaria* and bumblebees (like *Bombus impatiens, Bombus terrestris*), have been closely associated with pollination deficiencies in important crops like blueberries (*Vaccinium corymbosum*), tomatoes (*Solanum lycopersicum*) and almonds (*Prunus dulcis*). For example, in the United States, farms where wild populations of *Bombus* sp. have decreased as a result of habitat fragmentation and chemical exposure saw a significant decrease in blueberry yields [39]. According to Greenleaf and Kremen [29], tomato crops that rely on buzz pollination do well while native bumblebee numbers are steady, but they are badly served by honey bees. Research into incorporating wild pollinators for enhanced fruit set and crop stability has been conducted in California almond orchards due to the inadequacy of relying solely on *Apis mellifera* [43].

6.3 Mismatches in Timing

Reduced pollination efficiency results from phenological mismatches, which happen when plant blossoming and bee activity become asynchronous. A number of case studies illustrate the negative effects of these interruptions.

Case Study 1: Rocky Mountain Columbine (*Aquilegia coerulea*) and Bumblebees: Early flowering in *Aquilegia coerulea* in the Colorado Rocky Mountains may occur due to warmer temperatures. However, its main pollinators, bumblebees

(*Bombus balteatus*) emerge at their usual time of the year. In certain populations, this mismatch has resulted in a 50% decrease in seed output [25].

Case Study 2: Apple Orchards in the UK: According to a study conducted in British apple orchards, flowering progressed by 8 days over a 20-year period, while solitary bee emergence only moved by 2 days, resulting in a 6-day lag. This led to lower yields and poor fruit set [24].

Case Study 3: Pollination Deficit in Scandinavian Plants: In Scandinavian habitats, early-flowering species like bilberry (*Vaccinium myrtillus*) and bluebell (*Campanula rotundifolia*) have experienced earlier bloom due to climate warming. Meanwhile, the emergence of their main pollinators—including bumblebees (*Bombus pascuorum* and *Bombus terrestris*) and hoverflies—has not advanced proportionally. As a result, studies report a 20–30% decrease in seed set in these species compared to years with better synchrony [35].

6.3.1 Broader Consequences of Mismatches

A reduction in seed viability occurs when plants lack successful pollination which jeopardizes population sustainability [64].

Success in pollination disruptions disrupts both plant populations and ecological preferential cascades from herbivores to top predators because numerous animals count on pollinated plants for eating [51].

Farmers face annual agricultural losses between $235 billion and $577 billion because of pollinator shortage and pollination misalignments between crops and pollinators [38].

The stability of pollination networks suffers severe ecological together with agricultural consequences from pollination rhythm disturbances. Three factors resulting from climate change create problems for pollination networks by causing mismatches, pollinator declines and altered flowering times. Plants will have resilient relationships with pollinators only when we protect pollinator diversity and decrease pesticide applications and restore habitats [61].

7 Real-World Examples

The study of natural bee-plant relationships demonstrates how precise timing impacts pollination during agricultural farming and in untamed ecological settings and when ecosystems encounter environmental stressors. According to Garibaldi et al. [28], and Rader et al. [63], the relationships between bees and plants form a vital connection affecting agricultural output results and natural biodiversity as well as ecological systems' long-term balance.

7.1 Honeybees and Agricultural Crops

The species *Apis mellifera* serves as the primary pollinating agent for various commercial farms thus significantly enhancing worldwide food output. The pollinating capabilities of honeybees depend on their perfect matching of foraging patterns with crop blooming periods [43].

For instance, *A. mellifera* is the only commercial pollinator that provides almost all pollination services in Californian almond (*Prunus dulcis*) orchards, mobilizing over 2 million hives a year during peak bloom. Similar to this, *A. mellifera* is essential for boosting seed output and quality in canola (*Brassica napus*) and sunflower (*Helianthus annuus*) crops throughout Europe, particularly when colonies are positioned carefully to coincide with peak anthesis [19, 27].

Research has demonstrated that *A. mellifera* visitation significantly boosts fruit set and bean weight in coffee (*Coffea arabica*) plantations in tropical countries such as Brazil and Costa Rica. This is especially true when flowering is brief and necessitates prompt pollinator access [66]. The significance of synchrony in pollination services is demonstrated by the honeybee activity's compatibility with the floral rhythms of these crops.

The entire almond production in California relies exclusively on honeybees for pollination tasks. The flowering season of *Prunus dulcis* (almond trees) in early spring stimulates beekeepers to move millions of honeybee colonies for optimal pollination timing that promote the high nut and yield production [9].

Honeybees enhance seed production within sunflower (*Helianthus annuus*) and canola (*Brassica napus*) fields because they pollinate flowers when nectar is abundant. Scientific studies in Canada demonstrated that canola yields by 46% when bees carried out pollination tasks in comparison to wind pollination [55].

7.2 Bumblebees and Wildflowers

Bumblebees (*Bombus* sp.) play a crucial role in pollinating wild plants, especially native wildflowers. Unlike honeybees, they are well-adapted to working in cooler weather, low light and challenging environments, which makes them especially important for early-blooming and high-altitude plants [32, 36]. Thanks to their strong flight muscles, large size and long tongues, bumblebees can reach the nectar of deep, tube-shaped flowers like *Aconitum*, *Digitalis* and *Delphinium*—flowers that many other pollinators simply can't access. These wild plants have come to rely on bumblebees for successful pollination and over time, their blooming periods have become closely aligned with when bumblebees are active [17].

In cooler, mountainous, or temperate areas, it's essential that wildflowers bloom in sync with the emergence of their bumblebee pollinators. For example, species like *Bombus lucorum* and *Bombus pascuorum* begin foraging early in the spring, just as plants such as *Primula veris* and *Pulmonaria officinalis* come into bloom. However, climate change and habitat changes are now throwing off this timing. When bees and flowers fall out of sync, it can result in fewer pollination events, poor seed production and ultimately a decline in wildflower populations [48].

Bumblebees function as the essential pollinators for red clover plants (*Trifolium pratense*) since their floral nectar can only be reached by long-tongued pollinators. The decrease of bumblebee populations results 80% decrease in seed production in red clover [33].

Through regard for time, Bumblebees reach their peak activity rate during mid-morning since Foxglove (*Digitalis purpurea*) flowers release their maximum nectar supply. The timing of these activities results in successful pollen transfer and enables the flowers to exchange pollen between different plants [74].

Bumblebees within Arctic and Alpine ecosystems alter their strategy for habitat exploration to correspond with the brief flowering period. The Arctic poppy (*Papaver radicatum*) flowering cycle in Scandinavia triggers *Bombus polaris* to carry out its foraging activities which produces seeds in those harsh environmental conditions [49].

7.3 Impact of Environmental Stress on Pollination

Natural pollination frequencies become disrupted by environmental stressors including climate change together with habitat destruction and pesticide exposure thus causing plants to reproduce less and destabilizing ecosystems [60].

7.3.1 Climate Change and Timing Mismatches

Over the previous 50 years, Japanese scientists have documented how enhanced temperature result in early flowering of cherry trees (*Prunus serrulata*) by approximately 5–10 days. The delayed emergence timing of honeybees has created reduced fruit production [48].

Alpine (*Delphinium barbeyi*) flowers in the Rocky Mountains flower too early causing their bumblebee pollinators to miss out on pollination which leads to a 30–40% decreased seed production [25].

7.3.2 Pesticides and Pollinator Behaviour

The use of neonicotinoid pesticides has a noticeable impact on the natural foraging behaviour of honeybees. When bees come into contact with these chemicals—particularly imidacloprid—they tend to lose their usual foraging rhythm. A study by Henry et al. [37] found that bees exposed to this pesticide returned to their hives 55% less often, which ultimately reduced the colony's overall efficiency in collecting food and pollinating crops.

Similarly, heavy pesticide use on European farms has been linked to a decline in bumblebee (*Bombus terrestris*) populations. According to Gill et al. [31], neonicotinoid exposure disrupts bumblebees' ability to forage effectively, resulting in smaller, weaker colonies and fewer queens being produced. This decline in bee health also affects the plants they pollinate. For instance, wildflowers like *Centaurea nigra* (black knapweed) and *Papaver rhoeas* (common poppy) experienced up to a 30% drop in pollination success, largely due to reduced visits by bees and limited pollen transfer.

7.3.3 Habitat Loss and Pollen Deficits

Plant reproduction has been negatively impacted by the dramatic drop in native bee populations brought on by the fragmentation of Amazonian forests. For example, in regions where pollinators had vanished, fruit output for the Brazil nut tree (Bertholletia excelsa), which is highly dependent on particular local bee species for pollination, decreased by up to 80% [30].

Over 75% of native wildflower species in the UK rely on bees to pollinate them. The pollinator networks between plants and bees got disrupted due to urban de-elopement and habitat destruction during 1950s it resulted in reduction in number of wildflower by ~ 50% [6].

8 Conservation of Pollinator-Plant Rhythms

The most successful action for habitat conservation combines native plant reintroduction with pollinator-friendly landscape construction to facilitate a broad range of pollinating species [20]. The protection of nocturnal pollinators like moths and bats requires minimizing artificial light exposure because excessive artificial lights disrupt their night time foraging behaviour [59]. A fundamental conservation approach contains climate-resilient planting because it involves selecting plant species with wide phenological spans to protect floral resources from changing climatic conditions [64].

In agriculture, the conservation of pollinators depends heavily on sustainable farming methods. Improving pollinator populations and their pollination services can be achieved through limited pesticide application and combining crop rotation with flowering plants and setting up pollinator paths [61]. The involvement of citizens in scientific activities for pollinators and plant observations creates essential base for planning future conservation efforts. Table 1 offers a summary of the implemented strategies together with their positive effects on the environment.

Table 1 Conservation strategies for pollinator-plant rhythms and their ecological benefits

Conservation strategy	Implementation approach	Ecological benefits	References
Habitat Restoration	Reintroducing native flowering plant species, creating pollinator-friendly landscapes	Enhances pollinator diversity and stabilizes plant-pollinator interactions	[20]
Reduction of Artificial Light	Implementing dark-sky policies, reducing light pollution in key pollination zones	Supports nocturnal pollinators like moths and bats, preventing disorientation and behavioural changes	[59]
Climate-Resilient Planting	Selecting plant species with broad phenological ranges to adapt to changing climatic conditions	Ensures continuous floral resources, reducing phenological mismatches	[64]
Sustainable Agricultural Practices	Reducing pesticide use, incorporating pollinator-friendly crop rotations, establishing pollinator corridors	Maintains pollinator health, improves agricultural pollination services	[61]
Citizen Science & Pollinator Monitoring	Engaging local communities to track pollinator activity and flowering times	Provides valuable data for conservation planning and early detection of disruptions in pollination rhythms	[46]

8.1 *Maintaining Pollination Networks*

Ecosystem stability together with biodiversity depends on pollination networks since they support both plant species reproduction and pollinator survival. Expanding these networks needs protection of habitats and restoration work as well as the establishment of new connections [61].

Natural landscape protection of wildflower meadows plus forests as well as wetlands guards the diversity of pollinating species. The rhythm of pollination breaks when habitats get separated because plant populations become isolated from pollinator populations [77].

Natural and semi-natural habitat connections established between ecological corridors allow pollinators to move with ease between floral resources so they can maintain their synchronized foraging habits [45].

8.2 Pollinator-Friendly Practices

The specific techniques of land management together with agricultural practices create measurable changes for pollination patterns which influence bee survival rates while affecting plant ability to reproduce. Both plants and bees build their synchronized patterns through the implementation of pollinator-friendly approaches.

The use of pesticides such as neonicotinoids affects bee navigation systems and foraging activities along with causing overall negative impacts on colony health [34]. Organic farming and integrated pest management practices lower the chemical exposure affecting pollinators as well as minimizing pesticide mediated harm.

Agroforestry practices combined with wildflower strip management create refurbished landscapes which supply pollinators constant access to food across different seasons [28].

The combination of crop rotation with intercropping and minimal soil disturbance practices supports native pollinators while protecting soil biodiversity as per Klein et al. [43].

8.3 Protecting Pollinator Populations

Pollinator conservation relies on habitat protection coupled with passing new legislation and involving communities in its implementation.

Natural habitats such as tree hollows, soil burrows and deadwood need protection because they support wild bee species together with solitary bees and stingless bees [67].

Climate Change Mitigation patterns affect temperature and precipitation levels which cause flowers to bloom before pollinators start emerging. Conservation initiatives should focus on reforestation strategies combined with habitat linking and sustainable species selection [68].

Legislative bodies together with policy-making organizations continue to prioritize regulations about pesticide usage as well as habitat-defence initiatives and scientific studies of pollination ecology [38].

9 Conclusion

As essential biological timekeepers in ecosystems, bees support effective pollination procedures that support plant reproduction and overall ecological resilience. Their ability to synchronize foraging activity with floral phenology is a result of their evolutionary adaptations. This allows them to ensure successful seed set, improved genetic diversity and optimal fertilization, all of which support agricultural productivity and biodiversity [41].

The mutualistic relationship between flowering plants and their bee pollinators plays a fundamental role in maintaining species richness and promoting gene flow across plant populations. Disruption of these intricate interactions—due to anthropogenic drivers such as habitat loss, urbanization and climate-induced phenological mismatches—has been shown to impair reproductive success in numerous plant species, ultimately leading to biodiversity decline and ecological instability [56]. Bees also act as ecological keystone species, sustaining food web dynamics by contributing to fruit and seed production that benefits a wide range of organisms including frugivorous birds and herbivores [28].

However, this synchrony is increasingly under threat. Habitat degradation, pesticide exposure and climate variability have collectively weakened pollinator networks, leading to declines in pollination services that are vital for global food systems [61]. The long-term sustainability of ecosystems and food security thus hinges on urgent and integrated conservation responses, including pollinator-friendly land use, habitat restoration and the implementation of climate-smart agricultural practices [20].

Future strategies should leverage emerging insights from pollination biology, chronobiology and conservation genomics. Novel approaches such as assisted pollination, floral resource augmentation and selection for climate-resilient pollinator traits could play a pivotal role in safeguarding these interactions. Ultimately, preserving natural pollination rhythms is not only vital for ecosystem functioning and species conservation but also for ensuring the resilience of agroecosystems in the face of accelerating environmental change [43].

References

1. Albrecht M, Schmid B, Hautier Y, Müller CB (2012) Diverse pollinator communities enhance plant reproductive success. Proc R Soc B: Biol Sci 279(1748):4845–4852
2. Andrés F, Coupland G (2012) The genetic basis of flowering responses to seasonal cues. Nat Rev Genet 13(9):627–639
3. Ashman TL, Knight TM, Steets JA, Amarasekare P, Burd M, Campbell DR, Wilson WG et al (2004) Pollen limitation of plant reproduction: ecological and evolutionary causes and consequences. Ecology 85(9):2408–2421
4. Barrett SC, Harder LD (2006) David G. Lloyd and the evolution of floral biology: from natural history to strategic analysis. In: Ecology and evolution of flowers, pp 1–21
5. Bartomeus I, Ascher JS, Wagner D, Danforth BN, Colla S, Kornbluth S, Winfree R (2011) Climate-associated phenological advances in bee pollinators and bee-pollinated plants. Proc Natl Acad Sci 108(51):20645–20649
6. Biesmeijer JC, Roberts SP, Reemer M, Ohlemuller R, Edwards M, Peeters T, Kunin WE et al (2006) Parallel declines in pollinators and insect-pollinated plants in Britain and the Netherlands. Science 313(5785):351–354
7. Bloch G, Bar-Shai N, Cytter Y, Green R (2017) Time is honey: circadian clocks of bees and flowers and how their interactions may influence ecological communities. Philos Trans R Soc B: Biol Sci 372(1734):20160256
8. Bommarco R, Lundin O, Smith HG, Rundlöf M (2012) Drastic historic shifts in bumble-bee community composition in Sweden. Proc R Soc B: Biol Sci 279(1727):309–315
9. Brittain C, Kremen C, Klein AM (2013) Biodiversity buffers pollination from changes in environmental conditions. Glob Change Biol 19(2):540–547

10. Bronstein JL (1994) Our current understanding of mutualism. Quarter Rev Biol 69(1):31–51
11. Buchmann SL, Hurley JP (1978) A biophysical model for buzz pollination in angiosperms. J Theor Biol 72(4):639–657
12. Burkle LA, Alarcón R (2011) The future of plant–pollinator diversity: understanding interaction networks across time, space, and global change. Am J Bot 98(3):528–538
13. Burkle LA, Marlin JC, Knight TM (2013) Plant-pollinator interactions over 120 years: loss of species, co-occurrence, and function. Science 339(6127):1611–1615
14. Cardinal S, Danforth BN (2013) Bees diversified in the age of eudicots. Proc R Soc B: Biol Sci 280(1755):20122686
15. Chittka L, Menzel R (1992) The evolutionary adaptation of flower colours and the insect pollinators' colour vision. J Compar Physiol A 171:171–181
16. Chittka L, Thomson JD (eds) (2001) Cognitive ecology of pollination: animal behaviour and floral evolution. Cambridge University Press
17. Corbet SA, Williams IH, Osborne JL (1991) Bees and the pollination of crops and wild flowers in the European Community. Bee World 72(2):47–59
18. Danforth BN, Minckley RL, Neff JL, Fawcett F (2019) The solitary bees: biology, evolution, conservation. Princeton University Press, Princeton
19. Delaplane KS, Mayer DF (2000) Crop pollination by bees. CABI Publishing
20. Dicks LV, Breeze TD, Ngo HT, Senapathi D, An J, Aizen MA, Potts SG et al (2021) A global-scale expert assessment of drivers and risks associated with pollinator decline. Nat Ecol Evol 5(10):1453–1461
21. Dyer AG (2006) Bumblebees directly perceive variations in the spectral quality of illumination. J Compar Physiol A 192:333–338
22. Fenske MP, Imaizumi T (2016) Circadian rhythms in floral scent emission. Front Plant Sci 7:462
23. Fenster CB, Armbruster WS, Wilson P, Dudash MR, Thomson JD (2004) Pollination syndromes and floral specialization. Annu Rev Ecol Evol Syst 35(1):375–403
24. Fitter AH, Fitter RSR (2002) Rapid changes in flowering time in British plants. Science 296(5573):1689–1691
25. Forrest JR (2015) Plant–pollinator interactions and phenological change: what can we learn about climate impacts from experiments and observations? Oikos 124(1):4-13@@@
26. Forrest J, Miller-Rushing AJ (2010) Toward a synthetic understanding of the role of phenology in ecology and evolution. Philos Trans R Soc B: Biol Sci 365(1555):3101–3112
27. Free JB (1993) Insect pollination of crops
28. Garibaldi LA, Steffan-Dewenter I, Winfree R, Aizen MA, Bommarco R, Cunningham SA, Klein AM et al (2013) Wild pollinators enhance fruit set of crops regardless of honey bee abundance. Science 339(6127):1608–1611
29. Greenleaf SS, Kremen C (2006) Wild bees enhance honey bees' pollination of hybrid sunflower. Proc Natl Acad Sci 103(37):13890–13895
30. Giannini TC, Cordeiro GD, Freitas BM, Saraiva AM, Imperatriz-Fonseca VL (2015) The dependence of crops for pollinators and the economic value of pollination in Brazil. J Econ Entomol 108(3):849–857
31. Gill RJ, Ramos-Rodriguez O, Raine NE (2012) Combined pesticide exposure severely affects individual-and colony-level traits in bees. Nature 491(7422):105–108
32. Goulson D (2003) Bumblebees: their behaviour and ecology. Oxford University Press, USA
33. Goulson D, Darvill B (2004) Niche overlap and diet breadth in bumblebees; are rare species more specialized in their choice of flowers? Apidologie 35(1):55–63
34. Goulson D, Nicholls E, Botías C, Rotheray EL (2015) Bee declines driven by combined stress from parasites, pesticides, and lack of flowers. Science 347(6229):1255957
35. Hegland SJ, Nielsen A, Lázaro A, Bjerknes A-L, Totland Ø (2009) How does climate warming affect plant-pollinator interactions? Ecol Lett 12(2):184–195. https://doi.org/10.1111/j.1461-0248.2008.01269.x
36. Heinrich B (1979) Bumblebee economics. Harvard University Press, Cambridge

37. Henry M, Beguin M, Requier F, Rollin O, Odoux JF, Aupinel P, Decourtye A et al (2012) A common pesticide decreases foraging success and survival in honey bees. Science 336(6079):348–350
38. Intergovernmental Science-Policy Platform on Biodiversity and Ecosystem Services-IPBES (2016) Summary for policymakers of the thematic assessment on pollinators, pollination and food production. Biota Neotrop 16(1):e20160101
39. Isaacs R, Kirk AK (2010) Pollination services provided to small and large highbush blueberry fields by wild and managed bees. J Appl Ecol 47(4):841–849
40. Kerr JT, Pindar A, Galpern P, Packer L, Potts SG, Roberts SM, Pantoja A et al (2015) Climate change impacts on bumblebees converge across continents. Science 349(6244):177–180
41. Kevan PG, Imperatriz-Fonseca VL (2002) Pollinating bees: the conservation link between agriculture and nature
42. Klatt BK, Holzschuh A, Westphal C, Clough Y, Smit I, Pawelzik E, Tscharntke T (2014) Bee pollination improves crop quality, shelf life and commercial value. Proc R Soc B: Biol Sci 281(1775):20132440
43. Klein AM, Vaissière BE, Cane JH, Steffan-Dewenter I, Cunningham SA, Kremen C, Tscharntke T (2007) Importance of pollinators in changing landscapes for world crops. Proc R Soc B: Biol Sci 274(1608):303–313
44. Knauer AC, Kokko H, Schiestl FP (2021) Pollinator behaviour and resource limitation maintain honest floral signalling. Funct Ecol 35(11):2536–2549
45. Kremen C, Williams NM, Aizen MA, Gemmill-Herren B, LeBuhn G, Minckley R, Ricketts TH et al (2007) Pollination and other ecosystem services produced by mobile organisms: a conceptual framework for the effects of land-use change. Ecol Lett 10(4):299–314
46. Kremen C, Ullman KS, Thorp RW (2011) Evaluating the quality of citizen-scientist data on pollinator communities. Conserv Biol 25(3):607–617
47. Kudo G, Cooper EJ (2019) When spring ephemerals fail to meet pollinators: mechanism of phenological mismatch and its impact on plant reproduction. Proc R Soc B 286(1904):20190573
48. Kudo G, Ida TY (2013) Early onset of spring increases the phenological mismatch between plants and pollinators. Ecology 94(10):2311–2320
49. Lázaro A, Lundgren R, Totland Ø (2009) Co-flowering neighbors influence the diversity and identity of pollinator groups visiting plant species. Oikos 118(5):691–702
50. Memmott J (1999) The structure of a plant-pollinator food web. Ecol Lett 2(5):276–280
51. Memmott J, Craze PG, Waser NM, Price MV (2007) Global warming and the disruption of plant–pollinator interactions. Ecol Lett 10(8):710–717
52. Michener CD (2007) The bees of the world. JHU Press.
53. Moore D, Doherty P (2009) Acquisition of a time-memory in forager honey bees. J Compar Physiol A 195:741–751
54. Moore D, Rankin MA (1985) Circadian locomotor rhythms in individual honeybees. Physiol Entomol 10(2):191–197
55. Morandin LA, Winston ML (2005) Wild bee abundance and seed production in conventional, organic, and genetically modified canola. Ecol Appl 15(3):871–881
56. Ollerton J (2021) Pollinators and pollination: nature and society. Pelagic Publishing Ltd.
57. Ollerton J, Lack A (1998) Relationships between flowering phenology, plant size and reproductive success in shape Lotus corniculatus (Fabaceae). Plant Ecol 139:35–47
58. Ollerton J, Winfree R, Tarrant S (2011) How many flowering plants are pollinated by animals? Oikos 120(3):321–326
59. Owens AC, Lewis SM (2018) The impact of artificial light at night on nocturnal insects: a review and synthesis. Ecol Evol 8(22):11337–11358
60. Potts SG, Biesmeijer JC, Kremen C, Neumann P, Schweiger O, Kunin WE (2010) Global pollinator declines: trends, impacts and drivers. Trends Ecol Evol 25(6):345–353
61. Potts SG, Imperatriz-Fonseca V, Ngo HT, Aizen MA, Biesmeijer JC, Breeze TD, Vanbergen AJ et al (2016) Safeguarding pollinators and their values to human well-being. Nature 540(7632):220–229

62. Potts SG, Imperatriz-Fonseca V, Ngo HT, Biesmeijer JC, Breeze TD, Dicks LV, Vanbergen AJ et al (2016) The assessment report on pollinators, pollination and food production: summary for policymakers. In: Secretariat of the intergovernmental science-policy platform on biodiversity and ecosystem services
63. Rader R, Bartomeus I, Garibaldi LA, Garratt MP, Howlett BG, Winfree R, Woyciechowski M et al (2016) Non-bee insects are important contributors to global crop pollination. Proc Natl Acad Sci 113(1):146–151
64. Rafferty NE, CaraDonna PJ, Bronstein JL (2015) Phenological shifts and the fate of mutualisms. Oikos 124(1):14–21
65. Raguso RA, Willis MA (2002) Synergy between visual and olfactory cues in nectar feeding by naıve hawkmoths, Manduca sexta. Anim Behav 64(5):685–695
66. Ricketts TH, Daily GC, Ehrlich PR, Michener CD (2004) Economic value of tropical forest to coffee production. Proc Natl Acad Sci 101(34):12579–12582
67. Ricketts TH, Regetz J, Steffan-Dewenter I, Cunningham SA, Kremen C, Bogdanski A, Viana BF et al (2008) Landscape effects on crop pollination services: are there general patterns? Ecol Lett 11(5):499–515
68. Scaven VL, Rafferty NE (2013) Physiological effects of climate warming on flowering plants and insect pollinators and potential consequences for their interactions. Current Zool 59(3):418–426
69. Shemesh Y, Eban-Rothschild A, Cohen M, Bloch G (2010) Molecular dynamics and social regulation of context-dependent plasticity in the circadian clockwork of the honey bee. J Neurosci 30(37):12517–12525
70. Stone GN, Willmer P, Rowe JA (1998) Partitioning of pollinators during flowering in an African Acacia community. Ecology 79(8):2808–2827
71. Thomson JD (2010) Flowering phenology, fruiting success and progressive deterioration of pollination in an early-flowering geophyte. Philos Trans R Soc B: Biol Sci 365(1555):3187–3199
72. van Doorn WG, Van Meeteren U (2003) Flower opening and closure: a review. J Exp Botany 54(389):1801–1812
73. Willmer PG, Stone GN (2004) Behavioral, ecological, and physiological determinants of the activity patterns of bees. Adv Stud Behav 34(34):347–466
74. Willmer PG, Bataw AAM, Hughes JP (1994) The superiority of bumblebees to honeybees as pollinators: insect visits to raspberry flowers. Ecol Entomol 19(3):271–284
75. Willmer PG (1983) Thermal constraints on activity patterns in nectar-feeding insects
76. Willmer P (2011) Pollination and floral ecology. In: Pollination and floral ecology. Princeton University Press
77. Winfree R, Bartomeus I, Cariveau DP (2011) Native pollinators in anthropogenic habitats. Ann Rev Ecol Evolut Syst 42(1):1–22
78. Yon F, Joo Y, Cortés Llorca L, Rothe E, Baldwin IT, Kim SG (2016) Silencing Nicotiana attenuata LHY and ZTL alters circadian rhythms in flowers. New Phytol 209(3):1058–1066

MIX
Papier aus verantwortungsvollen Quellen
Paper from responsible sources
FSC® C105338

If you have any concerns about our products,
you can contact us on
ProductSafety@springernature.com

In case Publisher is established outside the EU,
the EU authorized representative is:
Springer Nature Customer Service Center GmbH
Europaplatz 3, 69115 Heidelberg, Germany

Printed by Libri Plureos GmbH
in Hamburg, Germany